Transhumanism and the Future of Humanity

Infinite Potential

TRANSHUMANISM AND THE FUTURE OF HUMANITY

Infinite Potential

F.H.

Transhumanism and the Future of Humanity

F.H.

ISBN 979 886 21 7344 4

BoFo YaY.

Contact: bofohaktaniyan@gmail.com

Preface: "Transhumanism and the Future of Humanity: Infinite Potential"

Throughout history, humanity has relentlessly pushed the boundaries of knowledge and progress. Evolving from primitive beings to the sophisticated species we are today, we have sought to conquer the challenges of our existence. Yet, our journey of self-improvement is far from over. As we stand on the threshold of a new era, a revolution of human potential calls us forward—the age of Transhumanism.

Welcome to "Transhumanism and the Future of Humanity: Infinite Potential," a compelling exploration of a groundbreaking field that delves into the realms of science, ethics, and the future of humanity. This book embarks on a captivating journey through the vast landscape of Transhumanism, delving deep into its philosophical roots and the ethical implications of human enhancement.

Transhumanism, as an interdisciplinary movement, seeks to harness the power of technology and biological advancements to transcend our physical and mental limitations. Its philosophical foundations revolve around the enhancement of human potential, the integration of technology with our biological selves, and the determination to overcome the barriers of mortality and the human condition.

Within these pages, we will explore the origins and evolution of Transhumanism, tracing its historical roots and understanding how this futuristic concept emerged as a prominent force shaping the future of humanity. We will delve into the intricate relationship between the human mind and technology, pondering the possibility of digital minds and consciousness transfer, as well as the potential ramifications of these revolutionary ideas.

Biological enhancement and longevity also find their place in our exploration, as we delve into the mechanisms of aging and the revolutionary anti-aging therapies that hold promise for a longer and healthier life. We shall also grapple with the ethical questions raised by genetic modifications and the transformation of human abilities into superhuman capacities through bioengineering.

As we venture further, we will immerse ourselves in the realms of virtual reality and digital imagination, discovering the impact of technology on our perception of reality and identity. We shall explore the concept of out-of-body experiences and the ethical dilemmas surrounding the notion of living in virtual worlds.

Throughout our journey, we will confront the ethical dimensions and societal implications of Transhumanism, reflecting on the ethical frameworks that guide our choices as a society and determine the course of our collective future.

"Transhumanism and the Future of Humanity: Infinite Potential" is an invitation to reflect on the potential of the human species and the ethical responsibilities that come with wielding the power of technology and biology. It is a call to contemplate the future we are forging and to consider the impact of our choices on generations yet to come.

I hope this book sparks curiosity and ignites a passion for the exploration of Transhumanism—a field that challenges our beliefs, expands our imagination, and pushes the boundaries of human understanding. Let us embark on this remarkable journey together, venturing into the uncharted territories of the human experience.

Table of Contents

CHAPTER 1

Introduction

1.1. Origins and Philosophical Foundations of Transhumanism

Transhumanism, an intriguing interdisciplinary movement, finds its roots in the profound desire of humanity to transcend its natural limitations and embrace a future where human potential knows no bounds. At its core, Transhumanism is built upon a philosophical foundation that seeks to enhance human capabilities through the integration of technology and biology.

The emergence of Transhumanism as a formal concept can be traced back to the mid-20th century when it first began to take shape. Over time, it has evolved and gained momentum, attracting thinkers, scientists, and visionaries from various disciplines. The term "Transhumanism" itself was coined by biologist Julian Huxley in 1957, capturing the essence of a movement that envisions the transformative potential of technology on human existence.

One of the fundamental principles of Transhumanism is the belief that human beings have the right to improve themselves and their circumstances through the application of science and technology. This philosophical underpinning asserts that humanity's evolutionary journey need not be bound by nature's constraints but can be guided by our own ingenuity and discoveries.

Within Transhumanist thought, the concept of "human enhancement" holds a central position. This refers to the use of technological interventions, such as genetic modifications, cognitive enhancements, and wearable devices, to augment human capabilities beyond their natural state. By embracing these enhancements, Transhumanists envision a future where humans can overcome physical and mental limitations and reach unprecedented levels of intellect, longevity, and well-being.

The movement also recognizes the potential of artificial intelligence and the integration of technology with the human mind. Ideas such as brain-computer interfaces and digital consciousness raise questions about the nature of human identity and the possibility of transferring human consciousness into digital realms.

However, Transhumanism is not without its ethical and societal challenges. As we strive for progress and improvement, ethical considerations surrounding issues of equality, access to technology, and the impact on social structures come to the forefront. The movement sparks debates about what it means to be human and the potential consequences of tampering with the fabric of human existence.

The origins and philosophical foundations of Transhumanism represent a daring exploration of the future of humanity. It is a movement that envisions a world where the synergy of technology and biology leads to an era of infinite potential, pushing the boundaries of human experience and what it means to be human. As the movement continues to evolve, it invites us to ponder the moral and philosophical implications of the choices we make on the path to shaping our own destinies.

1.1.1. The Emergence and Formation of the Transhumanism Concept

The concept of Transhumanism has its roots in a rich tapestry of ideas and influences that have shaped its emergence over time. Tracing its origins back to the mid-20th century, Transhumanism represents a visionary movement that seeks to elevate humanity to new heights through the integration of science, technology, and philosophy.

The seeds of Transhumanism were sown in the works of several prominent thinkers and writers who explored the potential of human transformation and the impact of technology on society. One of the earliest precursors to Transhumanist ideas can be found in the works of British biologist J.B.S. Haldane, who, in the 1920s, speculated about the possibility of "a new breed of beings" resulting from human cooperation with science and technology.

However, it was not until 1957 that the term "Transhumanism" was coined by Julian Huxley, a biologist and the first director of UNESCO. In his essay "Transhumanism," Huxley envisioned a future in which humanity could evolve beyond its current limitations through the application of science and technology. He saw a potential convergence between human biological and cultural evolution, envisioning a world where humans could consciously direct their own evolution.

During the 1960s and 1970s, further groundwork for the Transhumanism concept was laid by authors such as Olaf Stapledon, whose novel "Last and First Men" depicted the evolution of humanity into different species over the course of billions of years. Additionally, Robert Ettinger's book "The Prospect of Immortality," published in 1964, explored the concept of cryonics and the potential for future revival and rejuvenation of the human body.

As technological advancements accelerated in the latter half of the 20th century, the idea of Transhumanism gained more traction. In the 1980s, computer scientist and science fiction writer Vernor Vinge popularized the concept of the "Technological

Singularity," a hypothetical point in the future where artificial intelligence and technological progress would trigger radical changes in human civilization.

The 1990s marked a turning point for Transhumanism, as it began to coalesce into a more organized movement. The founding of the World Transhumanist Association (now Humanity+), by philosophers Nick Bostrom and David Pearce in 1998, provided a platform for like-minded individuals to advocate for the ethical use of technology to enhance human capabilities.

Transhumanism continued to gain momentum into the 21st century, attracting a diverse array of thinkers, scientists, and enthusiasts. With the rise of the internet and digital communication, the Transhumanist movement found new avenues for global collaboration and dissemination of ideas.

The philosophy of Transhumanism is characterized by a deep optimism about the future potential of humanity. It embraces a forward-looking and pro-technology stance, emphasizing the idea that human evolution can be guided and shaped by conscious choices, scientific understanding, and technological progress.

Critics of Transhumanism raise ethical concerns about potential risks and unintended consequences. The movement's rapid advancements in areas such as genetic engineering, artificial intelligence, and human-machine integration invite scrutiny over issues of social justice, equality, and the preservation of human dignity.

The emergence and formation of the Transhumanism concept represent a fascinating journey through the evolution of human thought and imagination. From the early musings of visionaries to the establishment of formal organizations, Transhumanism has become a thought-provoking movement that continues to inspire debate, challenge traditional beliefs, and explore the boundless possibilities of the human future. As we venture further into the 21st century, the ideals of Transhumanism will undoubtedly play a significant role in shaping the trajectory of human civilization.

1.1.2. Transhumanism's Key Philosophical Principles

Transhumanism is a forward-looking movement that seeks to enhance human capabilities through science and technology. Its core principles include advocating for responsible human enhancement, upholding individual autonomy and freedom, pursuing longevity and improved health, exploring the transformative potential of advanced technology, and engaging in ethical discussions about the societal impact of human enhancement. The movement promotes global collaboration to shape a future where humanity reaches new heights of potential and possibility.

1.1.2.1. Enhancement of Human Capabilities

The enhancement of human capabilities lies at the heart of Transhumanism, representing a visionary pursuit of elevating human potential beyond its natural limitations. This fundamental principle advocates the responsible and deliberate use of science and technology to augment various aspects of human existence.

At its core, the enhancement of human capabilities involves the application of scientific advancements to improve physical, cognitive, and emotional attributes. It seeks to overcome biological constraints, enabling individuals to lead healthier, more intellectually advanced, and fulfilling lives.

Physical Enhancement: Physical enhancement explores ways to overcome the limitations of the human body. This can encompass gene therapy, regenerative medicine, and advanced prosthetics. Genetic engineering may offer the potential to eradicate hereditary diseases and enhance physical traits, promoting strength, endurance, and longevity. Moreover, advancements in prosthetic limbs and exoskeletons empower those with disabilities to regain mobility and independence.

Cognitive Enhancement: Cognitive enhancement focuses on unlocking the full potential of the human mind. Techniques like brain-computer interfaces and neural implants aim to enhance memory, learning capabilities, and information processing.

Cognitive enhancement technologies may pave the way for accelerated learning, problem-solving, and creative thinking, revolutionizing education and cognitive development.

Emotional Enhancement: Emotional enhancement seeks to address mental well-being and emotional intelligence. Neuropharmacology and neuromodulation techniques may be utilized to treat mood disorders and promote emotional resilience. By better understanding and managing emotions, individuals could experience improved overall mental health and interpersonal relationships.

Despite the promises of human enhancement, ethical considerations loom large. Critics argue that these interventions might exacerbate social inequalities and lead to a "designer baby" phenomenon, where genetic selection influences human traits. Moreover, the potential risks and unforeseen consequences of altering human biology raise ethical dilemmas surrounding informed consent, individual identity, and societal norms.

Balancing the benefits and ethical concerns, Transhumanism advocates for responsible human enhancement, ensuring that scientific progress is accompanied by moral accountability. This includes robust regulatory frameworks, transparent public discourse, and an emphasis on individual autonomy. Additionally, promoting equal access to enhancement technologies is essential to prevent exacerbating existing societal disparities.

The enhancement of human capabilities represents a transformative vision that holds tremendous promise and challenges. As we navigate the frontiers of science and ethics, it is crucial to tread carefully, embracing the potential for human progress while maintaining our commitment to compassion, equality, and shared responsibility for shaping the future of humanity.

1.1.2.2. Integration of Technology and Humans

The integration of technology and humans is a pivotal concept within Transhumanism, representing the fusion of biological and technological elements to enhance human capabilities and redefine the human experience. This principle envisions a future where technology becomes an inseparable part of the human condition, leading to profound transformations in various aspects of life.

Biotechnology and Bioinformatics: The integration of technology with human biology encompasses biotechnology and bioinformatics. Advances in genetic engineering and gene editing offer the potential to cure hereditary diseases and enhance physical traits. Through bioinformatics, vast amounts of biological data can be analyzed and interpreted, leading to personalized medical treatments and preventive measures.

Brain-Computer Interfaces (BCIs): Brain-computer interfaces establish a direct link between the human brain and external devices, such as computers or prosthetics. BCIs enable communication, control, and information exchange through neural signals, empowering individuals with disabilities to interact with the world in new ways. The seamless interaction between the brain and technology opens doors to novel applications, from telepathic communication to controlling robotic systems with the power of thought.

Virtual and Augmented Reality: Virtual and augmented reality technologies offer immersive experiences that blur the boundaries between the physical and virtual worlds. By integrating technology into human perception, individuals can explore simulated environments, acquire new skills, and engage in interactive storytelling. Moreover, these technologies hold the potential to revolutionize education, training, and entertainment, enhancing learning and creativity.

Internet of Things (IoT) and Wearable Devices: The integration of technology into everyday life extends to the Internet of Things and wearable devices. IoT connects various devices, allowing seamless data sharing and automation. Wearable devices, such as smartwatches and fitness trackers, augment human capabilities by providing real-time health monitoring, activity tracking, and access to information on-the-go.

Challenges and Ethical Considerations: The integration of technology and humans raises significant challenges and ethical considerations. Privacy concerns arise as technology becomes deeply embedded in personal lives. Ensuring the security of personal data and protecting against unauthorized access becomes crucial. Additionally, reliance on technology may lead to issues like over-dependence, loss of privacy, and potential addiction to immersive virtual experiences.

Ethical dilemmas surrounding human autonomy and identity come to the forefront. Questions about the definition of humanity and the potential loss of individual agency in a technologically integrated world demand thoughtful exploration. Striking a balance between embracing technological advancements and preserving human values and dignity remains a complex task.

Embracing Responsible Integration: Transhumanism emphasizes responsible integration, wherein technological progress aligns with ethical considerations and human values. Transparent governance and inclusive decision-making processes are essential to navigate the implications of integrating technology and humans. Fostering a culture of open dialogue and public engagement can ensure that society collectively shapes the trajectory of technological advancements.

The integration of technology and humans represents a profound paradigm shift that holds transformative potential. As we embrace the opportunities offered by technology, it is vital to approach these advancements with ethical responsibility, ensuring that the fusion of human and technological elements enhances human well-being, preserves individual autonomy, and aligns with our collective vision of a prosperous and compassionate future.

1.1.2.3. Importance of Overcoming Limits and Advancement

At the core of Transhumanism lies a profound belief in the importance of overcoming human limits and advancing our potential as a species. This principle advocates for a future where humanity transcends its biological constraints through science, technology, and a collective commitment to progress. The significance of this pursuit can be understood through various lenses:

1. Expansion of Human Potential: Overcoming limits and advancement offer the possibility of expanding human potential beyond what was once thought possible. By pushing the boundaries of our physical, mental, and emotional capabilities, humanity can achieve feats that were once relegated to the realm of science fiction. This expansion opens doors to new opportunities, discoveries, and understandings of the universe and ourselves.

2. Medical and Scientific Breakthroughs: Advancing beyond current limitations holds the promise of groundbreaking medical and scientific breakthroughs. By conquering diseases, prolonging life, and enhancing human health, society can alleviate suffering and improve the quality of life for countless individuals. Moreover, scientific advancements resulting from this pursuit can have ripple effects across various fields, fueling innovation and progress.

3. Addressing Global Challenges: The quest to overcome limits and advance as a species is closely tied to addressing global challenges. From climate change to resource scarcity, the solutions to many pressing issues may lie in embracing scientific and technological progress. By harnessing our collective intelligence and capabilities, we can tackle some of humanity's most significant challenges and build a sustainable and thriving future.

4. Resilience and Adaptation: Overcoming limits fosters resilience and adaptability. As a species, our ability to adapt to changing circumstances is critical for survival and progress. Advancement allows us to evolve in response to evolving challenges, ensuring that we remain resilient in the face of adversity and uncertainty.

5. Fostering Curiosity and Exploration: The pursuit of overcoming limits fosters a culture of curiosity and exploration. It encourages us to question the status quo, seek new knowledge, and venture into uncharted territories. Embracing this curiosity-driven mindset sparks scientific inquiry and fuels the human spirit of exploration and discovery.

6. Empowerment and Human Agency: Advancing beyond limits empowers individuals and communities. It places the power of shaping the future in the hands of humanity, emphasizing human agency and choice. By embracing this principle, society can collectively take charge of its destiny and contribute to shaping a future that aligns with shared values and aspirations.

7. Ethical Responsibility and Global Collaboration: Overcoming limits and advancement underscore the ethical responsibility and need for global collaboration. As we embrace the transformative potential of science and technology, it becomes crucial to navigate these advancements responsibly, considering the implications on individuals, societies, and the planet. Global cooperation ensures that progress benefits all of humanity, fostering a more inclusive and equitable future.

The importance of overcoming limits and advancement in Transhumanism is profound and far-reaching. It encompasses expanding human potential, catalyzing medical breakthroughs, addressing global challenges, nurturing resilience and adaptability, fostering curiosity and exploration, empowering individuals, and recognizing the ethical responsibility of collective progress. By embracing this principle, humanity can embark on a transformative journey toward a future of greater possibilities, prosperity, and shared well-being.

1.1.3. The Historical Development and Evolution of Transhumanist Thought

Transhumanist thought emerged in the early 20th century, speculating about humanity's potential with science and technology. Coined in 1957 by Julian Huxley, the concept gained momentum in the 1990s with organized movements. Advancements in genetics and AI continue to shape Transhumanism, inspiring its ongoing evolution into the 21st century.

1.1.3.1. Early Precursors and Intellectual Movements of Transhumanism

The roots of Transhumanist thought can be traced back to early precursors and intellectual movements that explored the possibilities of human enhancement, scientific progress, and the convergence of biology and technology. Several key historical developments laid the groundwork for the emergence of Transhumanism:

1. Enlightenment and Rationalism: The Enlightenment era of the 17th and 18th centuries marked a significant intellectual movement that prioritized reason, science, and human progress. Philosophers such as René Descartes and Francis Bacon laid the groundwork for the scientific method, fostering an environment of critical inquiry and experimentation.

2. Utopian and Futurist Visions: In the 19th and early 20th centuries, various utopian and futurist visions emerged, imagining a future where technology and societal progress would lead to the betterment of humanity. Writers like H.G. Wells, Jules Verne, and Edward Bellamy depicted societies shaped by scientific and technological advancements.

3. Eugenics and Social Darwinism: During the late 19th and early 20th centuries, the ideologies of eugenics and social Darwinism gained prominence. These ideas posited that human heredity could be improved through selective breeding and that "survival of the fittest" principles should apply to human society. While controversial and often misused, these notions contributed to discussions about human evolution and improvement.

4. Russian Cosmism: In the late 19th and early 20th centuries, Russian Cosmism emerged as a philosophical and cultural movement. Its proponents, like Nikolai Fyodorov, advocated for the use of science and technology to overcome death and achieve human immortality. Cosmism laid the groundwork for discussions on human enhancement and the future of humanity.

5. Human Potential Movement: The Human Potential Movement of the 1960s and 1970s emphasized self-improvement and personal development through various psychological and spiritual practices. While not directly related to Transhumanism, this movement contributed to the broader exploration of human capabilities and the quest for self-transcendence.

6. The Cybernetics Movement: The cybernetics movement of the mid-20th century explored the concepts of feedback, communication, and control in both biological and technological systems. It paved the way for discussions on the integration of technology with the human body and mind.

7. Science Fiction and Futurism: Science fiction literature and futurist thinking have long envisioned societies where technology transforms human existence. Works of authors like Isaac Asimov, Arthur C. Clarke, and Ray Kurzweil popularized ideas such as artificial intelligence, mind uploading, and human-machine integration.

Together, these early precursors and intellectual movements provided a fertile ground for the emergence of

Transhumanism. They fostered discussions on human potential, scientific progress, and the possibilities of a future where technology could enhance human capabilities and redefine the human experience. As Transhumanism continues to evolve, its historical foundations remain an essential part of its philosophical landscape.

1.1.3.2. Transhumanism's Social Perception and Reputation

Transhumanism's social perception and reputation have been a subject of diverse views and interpretations. As a visionary movement seeking to enhance human capabilities through science and technology, it has elicited both fascination and skepticism from various segments of society.

Fascination and Optimism: Transhumanism has captivated individuals who embrace technological progress and the prospect of a future where human limitations can be overcome. Many proponents view it as a gateway to a utopian era, envisioning a world with extended lifespans, improved health, and enhanced cognitive abilities. Optimists see the movement as an opportunity to eradicate diseases, reduce suffering, and elevate humanity to new heights of achievement.

Skepticism and Ethical Concerns: Transhumanism has also faced skepticism and ethical concerns. Some critics worry about the potential consequences of human enhancement, fearing the creation of a "post-human" society with significant social disparities. Ethical dilemmas surrounding genetic engineering, mind uploading, and AI have raised questions about the preservation of individual identity, personal autonomy, and societal values.

Fear of Playing God: Another aspect of Transhumanism's social perception is the fear of "playing God." Critics argue that tampering with human biology and evolution could lead to unintended consequences and moral dilemmas. The fear of losing touch with humanity's intrinsic nature and reverence for life prompts caution and hesitation from various individuals and institutions.

Misunderstandings and Misrepresentations: Like any complex movement, Transhumanism has experienced misunderstandings and misrepresentations in popular media and public discourse. Exaggerations, sensationalism, and misconceptions have occasionally led to an oversimplification of its principles, contributing to polarized opinions.

Lack of Public Awareness: Transhumanism's social perception has also been affected by a lack of widespread public awareness. Despite gaining traction among certain intellectual circles and technology enthusiasts, many individuals remain unfamiliar with the movement's core principles and objectives.

Global and Cultural Variances: Transhumanism's reputation varies across different global and cultural contexts. Societies with a strong emphasis on traditional values or religious beliefs may exhibit greater resistance to the movement's ideas. Conversely, regions with a strong focus on scientific progress and technology may embrace Transhumanism more readily.

Overall, Transhumanism's social perception and reputation are shaped by a complex interplay of optimism, skepticism, ethical considerations, media portrayals, and cultural factors. As the movement continues to evolve, fostering open dialogue and responsible engagement with societal concerns is vital to shaping a future where humanity's collective aspirations for progress and well-being can be realized.

1.2. Human Interaction with Science and Technology

Science and technology have become integral to modern life, shaping society and offering unprecedented possibilities for progress and innovation. Striking a balance between their benefits and ethical considerations is vital for a sustainable future.

1.2.1. The Impact of Technological Advancements on Human Life

Technological advancements have profoundly impacted human life, revolutionizing how we live, work, and communicate. From the advent of the internet and smartphones to breakthroughs in medicine and automation, technology has brought about significant changes that have shaped the modern world.

Communication has been revolutionized by instant global connectivity, enabling seamless interactions and information exchange across borders. Access to vast amounts of knowledge and resources has become readily available at our fingertips, empowering individuals with unprecedented opportunities for learning and growth.

In healthcare, technological advancements have improved medical diagnosis, treatment, and patient care. Innovations in medical devices and procedures have extended life expectancy and enhanced overall well-being, contributing to a healthier population.

Automation and artificial intelligence have transformed industries, streamlining processes and increasing efficiency. While presenting challenges in terms of job displacement, technology has also created new opportunities and industries that demand different skill sets.

Despite its many benefits, the impact of technological advancements also raises concerns. Privacy, security, and ethical considerations regarding data usage and AI applications require careful navigation. Striking a balance between embracing progress and addressing potential risks is vital for ensuring technology remains a force for positive change in human life.

1.2.1.1. Revolutions in Communication and Information Technologies

The advancements in communication and information technologies have triggered transformative revolutions, reshaping how we interact, access information, and communicate with the world. These technological breakthroughs have had a profound impact on human life, revolutionizing various aspects of society:

1. Internet and World Wide Web: The internet and the World Wide Web have been revolutionary in connecting the global population. The internet's development and widespread adoption have enabled instant communication and information exchange, breaking down geographical barriers and fostering a truly interconnected world.

2. Mobile and Smart Devices: The proliferation of mobile and smart devices, such as smartphones and tablets, has empowered individuals with unparalleled access to information and services on-the-go. These portable devices have become essential tools for communication, productivity, and entertainment.

3. Social Media and Online Communities: Social media platforms have revolutionized how people connect and interact, enabling virtual communities to form around shared interests and experiences. Social media's impact on information dissemination, public discourse, and activism has been significant.

4. Digital Media and Content Consumption: The shift from traditional media to digital platforms has transformed how we consume content, from news and entertainment to educational resources. Online streaming services and digital publishing have revolutionized media consumption habits.

5. Cloud Computing and Data Storage: Cloud computing has revolutionized data storage and accessibility. Users can now store vast amounts of data remotely, facilitating collaboration, flexibility, and seamless access to information across devices.

6. E-commerce and Online Shopping: E-commerce platforms have revolutionized the way we shop, providing a convenient and efficient means of purchasing goods and services from anywhere in the world. Online marketplaces have expanded business opportunities and consumer choices.

7. Information Access and Knowledge Sharing: The internet has democratized access to information and knowledge, enabling individuals worldwide to access educational resources, research materials, and expertise from various fields.

8. Real-Time Communication: Real-time communication tools, such as instant messaging and video conferencing, have transformed how we interact professionally and socially. These tools bridge the gap between individuals, regardless of their geographical locations.

While these communication and information technologies have brought immense benefits, they also raise concerns related to privacy, misinformation, and digital divide. As these technologies continue to evolve, addressing these challenges becomes crucial to ensuring a positive and inclusive impact on human life.

1.2.1.2. Technological Advancements in Health and Medicine

Technological advancements in health and medicine have ushered in a new era of healthcare, enhancing diagnosis, treatment, and overall patient care. These breakthroughs have significantly impacted human life, leading to improved health outcomes and extending life expectancy:

1. Medical Imaging and Diagnostics: Technological advancements in medical imaging, such as MRI, CT scans, and ultrasound, have revolutionized diagnostic capabilities. These non-invasive imaging techniques allow doctors to visualize internal structures and identify abnormalities with precision.

2. Minimally Invasive Surgeries: Advancements in surgical techniques have led to minimally invasive procedures, reducing patient trauma and recovery time. Procedures performed through small incisions with the aid of robotics or endoscopy have become more common.

3. Precision Medicine and Personalized Treatments: The rise of precision medicine has enabled healthcare professionals to tailor treatments based on an individual's genetic makeup, lifestyle, and environmental factors. Personalized therapies offer more effective and targeted approaches to disease management.

4. Telemedicine and Remote Healthcare: Telemedicine technologies have facilitated remote healthcare services, enabling patients to consult with healthcare providers from their homes. This has improved access to medical care, especially for individuals in remote or underserved areas.

5. Wearable Health Devices: Wearable health devices, such as fitness trackers and smartwatches, allow individuals to monitor their health in real-time. These devices track vital signs, physical activity, and sleep patterns, promoting preventive health measures.

6. Gene Editing and Gene Therapies: The development of gene editing technologies, like CRISPR-Cas9, holds promising implications for treating genetic disorders and modifying disease-causing genes. Gene therapies offer potential cures for previously untreatable conditions.

7. Digital Health Records and Data Analytics: Digital health records and data analytics have streamlined healthcare information management. Electronic records enable secure and efficient sharing of patient data, supporting informed decision-making and research.

8. Vaccine Development and Immunotherapy: Technological advancements have expedited vaccine development processes, as seen in the COVID-19 pandemic response. Immunotherapies, such as CAR-T cell therapy, have shown promise in cancer treatment by harnessing the body's immune system.

While technological advancements in health and medicine have brought remarkable progress, ethical considerations, data security, and accessibility remain crucial issues. Striking a balance between innovation, ethical standards, and equitable access to healthcare will continue to shape the future of medical technology's impact on human life.

1.2.1.3. Transportation and the Role of Technology

Technology has played a transformative role in revolutionizing transportation, shaping the way people and goods move across the globe. These technological advancements have had a profound impact on human life, improving mobility, efficiency, and sustainability:

1. Automobile and Aviation Innovations: Advancements in automobile and aviation technology have led to faster, safer, and more comfortable travel. From the development of electric and autonomous vehicles to the introduction of supersonic and hypersonic aircraft, technology has pushed the boundaries of transportation.

2. High-Speed Rail and Maglev Systems: High-speed rail and magnetic levitation (maglev) systems have transformed intercity and intracity transportation. These systems offer rapid transit options with reduced travel times and increased capacity.

3. Ride-Sharing and Mobility Services: The emergence of ride-sharing and mobility services has revolutionized urban transportation. Apps and platforms connecting passengers with drivers have become an integral part of modern urban mobility.

4. Smart Traffic Management: Smart traffic management technologies, including adaptive traffic signals and real-time traffic data, optimize traffic flow and reduce congestion in urban areas.

5. Sustainable Transportation Solutions: Technology has facilitated the development of sustainable transportation solutions, such as electric vehicles, hydrogen-powered vehicles, and biofuels. These technologies aim to reduce greenhouse gas emissions and combat climate change.

6. Drones and Delivery Services: The use of drones in transportation and delivery services has introduced new possibilities for logistics and last-mile deliveries, particularly in remote or challenging terrains.

7. Hyperloop and Future Transportation Concepts: Innovative transportation concepts like the Hyperloop propose high-speed, low-friction travel in near-vacuum tubes. These futuristic ideas could revolutionize long-distance transportation.

8. Connectivity and Smart Infrastructure: The integration of technology with transportation infrastructure has improved connectivity and efficiency. Smart traffic lights, digital toll systems, and real-time navigation apps streamline travel experiences.

9. Space Exploration and Commercial Spaceflight: Advancements in space exploration and commercial spaceflight are opening up new possibilities for human space travel and the potential colonization of other planets.

While technology has brought about significant improvements in transportation, challenges such as energy consumption, infrastructure development, and safety considerations remain important areas of focus. Striving for sustainable and inclusive transportation solutions will be essential to harnessing the full potential of technology in shaping the future of transportation and its impact on human life.

1.2.2. The Conceptual Transformation of Human Physical and Mental Boundaries

Advancements in science and technology have led to a conceptual transformation of human physical and mental boundaries. Through human enhancement, prosthetics, brain-computer interfaces, and cognitive technologies, the traditional limits of human capabilities are being redefined. This transformation challenges conventional notions of what it means to be human, opening new possibilities for human potential and exploring uncharted territories in the realms of physical and mental abilities.

1.2.2.1. Overcoming Physical Limitations and Experiencing Beyond the Body

Technological advancements have ushered in a new era where humans can overcome physical limitations and experience beyond the boundaries of their bodies. This transformative concept is achieved through various means:

1. Prosthetics and Augmentation: Advanced prosthetic limbs and exoskeletons enable individuals with physical disabilities to regain mobility and independence. These technologies bridge the gap between biological limitations and enhanced physical capabilities.

2. Human Enhancement: Human enhancement technologies, such as gene editing and biotechnology, offer the potential to improve physical traits and extend human lifespan. By transcending biological constraints, humans can achieve heightened physical capabilities and even potentially reverse the effects of aging.

3. Brain-Computer Interfaces (BCIs): BCIs establish a direct link between the human brain and external devices. They allow individuals to control computers, robotic limbs, and even communicate with others through thought alone, enabling experiences beyond the limitations of the physical body.

4. Virtual Reality (VR) and Augmented Reality (AR): VR and AR technologies create immersive experiences, transporting users to virtual worlds and augmenting their perception of reality. Through these technologies, individuals can explore new environments and scenarios, transcending the physical constraints of their immediate surroundings.

5. Digital Consciousness and Mind-Uploading Speculations: Transhumanist thinkers explore the concept of transferring human consciousness into digital formats. Although still speculative, the idea of mind-uploading suggests the potential for humans to experience existence beyond their physical bodies.

These developments challenge the traditional notions of what it means to be human and raise profound ethical and philosophical questions. As humans venture into a world where physical limitations can be surpassed and experiences extend beyond the body, it becomes essential to navigate the implications responsibly, considering individual autonomy, identity, and the impact on society at large.

1.2.2.2. Brain-Computer Interfaces and Digital Imagination

Brain-Computer Interfaces (BCIs) have emerged as groundbreaking technologies that facilitate direct communication between the human brain and external devices, such as computers and prosthetics. This innovative interface holds immense potential for transforming human experiences and unlocking new realms of digital imagination:

1. How Brain-Computer Interfaces Work: BCIs use sensors to detect brain activity, such as neural signals or electroencephalogram (EEG) patterns. These signals are then decoded and translated into commands that control external devices. This bidirectional communication allows users to interact with technology merely through their thoughts.

2. Enhancing Accessibility and Mobility: BCIs offer tremendous promise in enhancing accessibility and mobility for individuals with disabilities. People with motor impairments can use BCIs to control assistive technologies like robotic limbs or communicate through speech-generating devices, liberating them from physical limitations.

3. Virtual and Augmented Reality Experiences: BCIs enable immersive virtual and augmented reality experiences. Users can navigate virtual environments, manipulate objects, and interact with digital content merely through their brain activity, blurring the lines between the physical and virtual worlds.

4. Advancements in Gaming and Entertainment: BCIs have the potential to revolutionize the gaming and entertainment industry. Players can experience video games by using their thoughts, creating new dimensions of interaction and engagement.

5. Digital Creativity and Expression: The fusion of BCIs with digital creative tools allows artists to translate their thoughts and emotions into tangible digital art forms, pushing the boundaries of human imagination and expression.

6. Mind-Controlled Devices and Robotics: BCIs are instrumental in developing mind-controlled devices and robotics. This has applications in various fields, such as remote robotic control, prosthetics, and industrial automation.

7. Ethical and Privacy Considerations: As BCIs become more sophisticated, ethical and privacy concerns arise. Issues such as data security, informed consent, and potential misuse of brain data necessitate careful consideration and robust regulatory frameworks.

8. The Future of Digital Imagination: The ongoing development of BCIs paves the way for a future where humans can seamlessly interact with technology through their thoughts, expanding digital imagination beyond current horizons. However, ethical guidelines, transparency, and societal discourse are crucial to ensure that BCIs are developed and utilized responsibly.

Brain-Computer Interfaces and their integration with digital imagination offer transformative opportunities for human experiences. As we venture into this uncharted territory, it is essential to strike a balance between technological advancement and ethical considerations, shaping a future where BCIs enrich human lives and empower individuals while upholding fundamental values of privacy, autonomy, and human dignity.

1.2.2.3. The Digitalization of Human Identity and Avatars

The rapid advancement of technology has led to the digitalization of human identity, giving rise to the concept of avatars as digital representations of individuals. This transformative phenomenon has significant implications for how humans interact, perceive themselves, and engage in virtual spaces:

1. Virtual Identities and Avatars: Avatars are digital representations of individuals in virtual environments, gaming platforms, and social media. Users can customize their avatars to reflect their physical appearance, personality, and preferences.

2. Social Media and Online Persona: Social media platforms have become spaces where individuals curate and project their digital identities through posts, photos, and interactions. These online personas may differ from one's offline self, shaping how others perceive and interact with them.

3. Virtual Reality and Immersive Environments: In virtual reality (VR) and immersive environments, avatars play a central role in facilitating user interactions. Users can experience virtual worlds, communicate, and collaborate through their avatars, blurring the boundaries between the physical and digital realms.

4. Psychological Impact and Identity Exploration: The digitalization of human identity allows for novel forms of self-expression and identity exploration. Individuals may use avatars to experiment with different aspects of their identity, fostering self-discovery and personal growth.

5. Empathy and Social Connection: Avatars have the potential to enhance empathy and social connection. In virtual spaces, individuals interact with others based on their avatar's actions, emotions, and expressions, promoting a deeper understanding of diverse perspectives.

6. Privacy and Security Concerns: As avatars become an integral part of online interactions, privacy and security concerns arise. Users must navigate issues of data protection, identity theft, and the boundaries between their virtual and real-world personas.

7. Professional Avatars and Virtual Collaboration: In professional settings, avatars can represent users during virtual meetings and collaborative workspaces. This facilitates seamless global collaboration and reduces the need for physical presence.

8. Ethical Considerations and Digital Personhood: The digitalization of human identity raises ethical questions about digital personhood, ownership of digital representations, and the potential for identity manipulation and misrepresentation.

9. Potential for Escapism and Disconnection: While avatars offer exciting possibilities, they also present challenges related to potential escapism and disconnection from reality. Excessive reliance on virtual identities may impact real-world relationships and experiences.

As the digitalization of human identity and the use of avatars become more prevalent, society must navigate the complex implications thoughtfully. Ethical guidelines, transparency, and ongoing dialogue will be crucial in shaping a future where avatars enrich human experiences while preserving individual agency, privacy, and authenticity in both physical and digital realms.

1.2.3. The Societal and Ethical Implications of Transhumanism

Transhumanism presents profound societal and ethical implications. Its potential to enhance human capabilities, extend lifespan, and redefine human existence raises questions about identity, equality, and the distribution of benefits. Balancing progress with ethical considerations will be essential for a responsible and equitable integration of Transhumanist ideas into society.

1.2.3.1. Inequality and Social Justice Issues

Transhumanism's advancements may exacerbate existing societal inequalities and raise concerns related to social justice:

1. Access to Enhancements: Transhumanist technologies and treatments might initially be available only to the affluent, widening the gap between the enhanced and the non-enhanced, potentially creating a new form of privilege.

2. Economic Disparities: If certain enhancements become essential for economic competitiveness, those unable to afford or access them may face limited job opportunities and reduced social mobility.

3. Health Disparities: As medical enhancements become prevalent, those without access may experience reduced health outcomes, widening the health disparities gap.

4. Ethical Dilemmas in Enhancement: Decisions surrounding the use of enhancements raise ethical questions about consent, safety, and the potential for coercion or exploitation.

5. Identity and Authenticity: The blurring of physical and digital identities may challenge the authenticity of human experiences and relationships, leading to questions about true human connection.

6. Human Rights: Transhumanist interventions might challenge traditional notions of human rights, leading to discussions about the definition and protection of these rights.

Addressing these concerns requires proactive measures, such as establishing ethical guidelines, ensuring equitable access to enhancements, and fostering inclusive discussions about the impact of Transhumanism on society. By approaching these issues thoughtfully, society can navigate the implications of Transhumanism in a way that upholds principles of justice and fairness for all.

1.2.3.2. Ethical Values and the Interface with Traditional Beliefs

Transhumanism's rapid advancement raises intricate ethical questions and interfaces with traditional beliefs, sparking debates on how to reconcile technological progress with deeply rooted cultural and moral values:

1. Human Dignity and Nature: Traditional beliefs may emphasize the inherent dignity and sanctity of human life as it exists naturally. Transhumanism challenges these beliefs by proposing enhancements that alter human biology and potentially redefine what it means to be human.

2. Playing God and Hubris: Some religious and cultural perspectives caution against "playing God" and exceeding natural boundaries. Transhumanist pursuits of human enhancement may be viewed as an act of hubris, raising concerns about the consequences of tampering with nature.

3. Equality and Fairness: Ethical values of equality and fairness may clash with potential inequalities arising from access to enhancements. Ensuring equitable access to enhancements becomes essential to uphold these values.

4. Identity and Authenticity: Transhumanist technologies may blur the lines between physical and digital identities, leading to questions about personal authenticity and the integrity of human experiences.

5. Preservation of Culture and Tradition: Rapid technological changes can disrupt cultural norms and traditions, prompting discussions on preserving cultural heritage amid a rapidly changing world.

6. Autonomy and Consent: Ethical considerations include the importance of informed consent and personal autonomy in decisions related to human enhancements.

7. Unintended Consequences: Cultural and ethical values urge caution about potential unintended consequences of embracing certain technological advancements, considering their impact on societal values and norms.

8. Responsibility and Regulation: Striking a balance between technological progress and ethical values calls for responsible development and effective regulation of Transhumanist technologies.

Navigating the interface between Transhumanism and traditional beliefs requires open dialogue, collaboration, and sensitivity to diverse cultural perspectives. Integrating ethical values with technological innovation ensures that society benefits from Transhumanist advancements while safeguarding core human principles and beliefs. By fostering a holistic and inclusive approach, we can forge a path that respects the diversity of perspectives and leads to a future that integrates technological progress with ethical integrity.

CHAPTER 2

Mind Enhancement and Brain-Computer Interfaces

2.1. The Relationship Between Brain and Mind: Physical and Digital Mind Concepts

The relationship between the brain and mind is a complex and fundamental aspect of human cognition. Traditionally, the mind is considered an emergent property of the brain's neural activities, encompassing thoughts, emotions, and consciousness. However, emerging technologies, such as brain-computer interfaces, have sparked discussions about the possibility of a digital mind—a consciousness or identity that exists in virtual or digital environments. The exploration of these physical and digital mind concepts raises profound questions about human consciousness, identity, and the boundaries between the physical and digital realms.

2.1.1. Understanding the Relationship Between Brain and Mind

The relationship between the brain and mind is a complex and still not fully understood phenomenon. The mind is believed to emerge from the brain's intricate neural networks and processes, encompassing consciousness, thoughts, emotions, and perceptions. While scientific research sheds light on this connection, many aspects of how the brain gives rise to the mind remain a fascinating and ongoing exploration in the fields of neuroscience and cognitive science.

2.1.1.1. Neuroscience and Cognitive Neuroscience

Neuroscience: Neuroscience is a multidisciplinary field that explores the structure and function of the nervous system, including the brain, spinal cord, and peripheral nerves. It seeks to understand how these biological systems give rise to behavior, cognition, and the functioning of the mind. Neuroscience employs various methods, including neuroimaging, electrophysiology, and molecular biology, to study the brain's cellular and molecular processes, neural networks, and the neural basis of perception, memory, emotions, and other cognitive functions.

Cognitive Neuroscience: Cognitive neuroscience is a subfield of neuroscience that focuses on the neural basis of cognitive processes and mental functions. It aims to understand how the brain supports complex cognitive functions like attention, language, decision-making, learning, and problem-solving. By integrating techniques from neuroscience and cognitive psychology, cognitive neuroscience investigates how the brain and mind interact and how cognitive processes are localized and interconnected within the brain.

Neural Correlates of the Mind: Cognitive neuroscience research has identified neural correlates of various mental processes. For example, studies using functional magnetic resonance imaging (fMRI) have linked specific brain regions to specific cognitive functions. For instance, the prefrontal cortex is associated with decision-making, while the hippocampus is crucial for memory formation. Understanding these neural correlates helps reveal how the brain generates mental states and cognitive abilities.

Neuroplasticity: One essential discovery in neuroscience is neuroplasticity, which refers to the brain's ability to adapt and reorganize itself in response to experiences, learning, and injuries. This concept underpins our understanding of brain development, learning, and recovery from brain injuries.

Clinical Applications: Neuroscience and cognitive neuroscience have significant clinical applications. They provide insights into neurological disorders, psychiatric conditions, and cognitive deficits. Researchers and clinicians use this knowledge to develop therapies, treatments, and interventions to address various brain-related issues.

Ethical Considerations: Advancements in neuroscience and cognitive neuroscience raise ethical considerations, particularly concerning brain research, cognitive enhancement, and privacy. As our understanding of the brain-mind relationship grows, responsible ethical guidelines become essential for the ethical application of these insights.

In summary, neuroscience and cognitive neuroscience play a pivotal role in advancing our understanding of the relationship between the brain and mind. Their research sheds light on the neural underpinnings of cognition, emotions, and behavior, contributing to both scientific knowledge and potential clinical applications.

2.1.1.2. The Nature of Consciousness and the Mind-Body Problem

The nature of consciousness and the mind-body problem is one of the most profound and enduring mysteries in philosophy and neuroscience. This complex philosophical dilemma explores the relationship between the physical brain and subjective experiences of consciousness:

1. Consciousness Defined: Consciousness refers to our subjective experience of awareness and self-identity. It encompasses thoughts, perceptions, emotions, and the sense of being present in the world.

2. The Mind-Body Problem: The mind-body problem questions how subjective consciousness emerges from physical brain processes. It challenges us to reconcile the seemingly immaterial and private nature of consciousness with the physical and objective reality of the brain.

3. Philosophical Dualism: Some philosophical views propose dualism, positing that the mind and brain are fundamentally separate entities. René Descartes, for instance, argued for a dualistic perspective, suggesting that the mind (or soul) interacts with the body through the pineal gland.

4. Materialism and Physicalism: Materialism and physicalism assert that all phenomena, including consciousness, arise from physical processes. According to this view, the brain's neural activities generate consciousness, and there is no need to postulate a separate mind or soul.

5. Emergentism: Emergentism suggests that consciousness emerges as a result of complex interactions among brain neurons and their networks. In this view, consciousness is an emergent property arising from the organization and complexity of neural processes.

6. The Hard Problem of Consciousness: Philosopher David Chalmers coined the term "the hard problem of consciousness" to describe the challenge of explaining how subjective experiences arise from brain activities. This problem concerns why and how certain brain processes give rise to subjective consciousness.

7. Neural Correlates of Consciousness (NCC): Neuroscientists seek to identify the neural correlates of consciousness (NCC) - specific brain activities that correlate with conscious experiences. While progress has been made in this area, the question of how neural processes give rise to subjective experience remains unresolved.

8. Qualia and Subjectivity: Qualia are the subjective qualities of conscious experiences, such as the redness of a red apple or the feeling of pain. Understanding how these subjective experiences arise from physical brain processes remains a central challenge.

The nature of consciousness and the mind-body problem continue to inspire philosophical, scientific, and interdisciplinary investigations. Bridging the gap between the physical brain and the immaterial realm of consciousness remains a profound challenge, with profound implications for our understanding of human nature and the essence of subjective experience.

2.1.2. The Mental Abilities and Limitations of Artificial Intelligence

Artificial Intelligence (AI) excels in pattern recognition, problem-solving, and speed. It learns from data and handles repetitive tasks efficiently. However, AI lacks human-like creativity, emotional intelligence, and common sense reasoning.

Combining AI's strengths with human cognition can lead to more effective and responsible applications.

2.1.2.1. The History and Evolution of Artificial Intelligence

Artificial Intelligence (AI) has a rich history of development, marked by key milestones and breakthroughs:

1. Early Concepts (1950s-1960s): The field of AI emerged in the 1950s, with pioneers like Alan Turing proposing the Turing Test to determine a machine's ability to exhibit intelligent behavior. In the late 1950s and 1960s, researchers developed early AI programs to solve mathematical problems and play games like chess.

2. Symbolic AI (1960s-1970s): Symbolic AI, also known as Good Old Fashioned AI (GOFAI), focused on representing knowledge and rules using symbols and logic. The development of expert systems, which emulated human expertise in specific domains, was a significant advancement during this period.

3. AI Winter (1970s-1980s): Progress in AI faced challenges, and a period known as "AI Winter" ensued due to high expectations not meeting reality. Funding and interest in AI research declined during this time.

4. Neural Networks Resurgence (1980s-1990s): In the 1980s, neural networks experienced a resurgence, exploring parallel processing and learning algorithms inspired by the human brain. This led to significant developments in pattern recognition and machine learning.

5. Rise of Machine Learning (1990s-2000s): The 1990s saw the rise of machine learning algorithms, such as support vector machines and decision trees. AI applications expanded into areas like data mining, natural language processing, and computer vision.

6. Big Data and Deep Learning (2010s): The 2010s witnessed a revolution in AI, driven by the availability of big data and advances in computing power. Deep learning, a subset of machine learning using neural networks with multiple layers, achieved breakthroughs in image and speech recognition, leading to transformative applications.

7. AI Integration in Everyday Life (Present): AI has become pervasive in daily life, powering virtual assistants, recommendation systems, and autonomous vehicles. AI is also applied in healthcare, finance, and various industries, transforming how we live and work.

8. Ethical and Societal Considerations: The rapid evolution of AI has raised ethical and societal considerations.

Discussions on data privacy, bias, transparency, and responsible AI deployment have become essential in AI development.

As AI continues to evolve, ongoing research and development hold the potential to unlock new frontiers and address current limitations. Ethical and responsible practices will be paramount to ensure AI's benefits are harnessed while mitigating potential risks.

2.1.2.2. Artificial Intelligence's Sensory Perception and Cognitive Capacity

Sensory Perception: Artificial Intelligence exhibits remarkable progress in sensory perception, enabling machines to perceive and interpret the world through various senses:

Computer Vision: AI-powered computer vision systems can analyze images and videos, identifying objects, people, and activities with high accuracy. These systems find applications in surveillance, autonomous vehicles, and medical imaging.

Speech Recognition: AI enables machines to understand and transcribe human speech, leading to advancements in virtual assistants, speech-to-text technologies, and voice-controlled devices.

Natural Language Processing (NLP): NLP empowers AI to understand, interpret, and generate human language. Chatbots, language translation, and sentiment analysis are examples of NLP applications.

Tactile and Haptic Perception: Advancements in robotics and AI enable machines to sense and respond to touch and pressure, allowing for applications in robotics, prosthetics, and virtual reality.

Cognitive Capacity: AI demonstrates impressive cognitive capabilities that have led to breakthroughs in problem-solving and decision-making:

Machine Learning: AI's ability to learn from data allows it to recognize patterns, make predictions, and improve performance on tasks, surpassing human accuracy in some domains.

Deep Learning: Deep learning, a subset of machine learning, employs neural networks with multiple layers to perform complex tasks, such as image and speech recognition, natural language processing, and game playing.

Pattern Recognition: AI excels at identifying patterns in large datasets, enabling it to identify trends, anomalies, and correlations in various applications.

Big Data Processing: AI's computational capacity enables it to handle massive datasets, extracting valuable insights from big data and driving data-driven decision-making.

Simulation and Prediction: AI can simulate complex scenarios, predict outcomes, and model various systems, aiding in weather forecasting, economic predictions, and scientific simulations.

While AI's sensory perception and cognitive capacity continue to improve, it is essential to address challenges related to interpretability, bias, and transparency. As AI becomes more capable, responsible and ethical use becomes paramount to ensure that AI technologies enhance human experiences while addressing potential risks and societal concerns.

2.1.3. The Concept of Digital Minds and Discussions on Mind Transfer

The concept of digital minds revolves around the idea of creating consciousness or identity in digital or virtual environments. Discussions on mind transfer contemplate the possibility of transferring a person's consciousness or mind into a digital format. These ideas provoke profound ethical, philosophical, and scientific debates about the nature of consciousness, identity, and the boundaries between physical and digital existence. While still speculative, these discussions explore potential future scenarios that challenge our understanding of what it means to be human and the fundamental nature of consciousness itself.

2.1.3.1. Digital Minds and the Notion of Internet Consciousness

The notion of digital minds and the idea of internet consciousness explore the possibility of creating or hosting consciousness within digital environments, particularly the internet. While this concept remains speculative and highly debated, it raises thought-provoking questions about the nature of consciousness, identity, and the potential extension of human experiences in the digital realm.

1. Digital Minds and Artificial Consciousness: The idea of digital minds revolves around the possibility of creating conscious entities in digital or virtual spaces. This concept often intersects with the field of artificial consciousness, where researchers explore whether machines or AI systems can achieve consciousness and subjective experience. The development of sophisticated AI algorithms, brain-computer interfaces, and neural networks has sparked discussions about the potential for digital entities to exhibit consciousness-like qualities.

2. Internet Consciousness and Global Connectivity: The notion of internet consciousness speculates on the idea that the vast interconnectedness of the internet could give rise to a collective or global form of consciousness. Some proponents argue that the constant flow of information, ideas, and interactions

across the internet could create a distributed form of awareness, where humanity becomes part of a larger collective mind.

3. Ethical and Philosophical Considerations: Discussions on digital minds and internet consciousness raise profound ethical and philosophical questions. Ethical concerns focus on the implications of creating conscious entities in digital form, considering their rights, well-being, and potential for exploitation. Philosophical inquiries explore the nature of consciousness itself, the relationship between physical and digital existence, and the concept of identity in the digital age.

4. Technological and Scientific Challenges: The realization of digital minds or internet consciousness faces significant technological and scientific hurdles. Understanding the nature of consciousness and replicating it in a digital medium is a complex and unsolved problem. Additionally, ensuring the authenticity and fidelity of any digital consciousness raises technical challenges in terms of data processing, storage, and security.

5. Cultural and Societal Implications: The concept of internet consciousness raises questions about cultural identity, global connectivity, and the blurring of boundaries between individuals and the digital realm. It challenges conventional notions of individuality, selfhood, and the human experience.

While the notion of digital minds and internet consciousness remains speculative, it serves as a fertile ground for exploration and debate at the intersection of technology, ethics, and philosophy. As these discussions continue, society must approach these concepts with responsible, thoughtful consideration, reflecting on the potential impact on our understanding of consciousness, identity, and the future of humanity in an increasingly connected and digital world.

2.1.3.2. Mind Transfer and the Problem of Copying

The concept of mind transfer, also known as mind uploading or whole brain emulation, envisions transferring an individual's consciousness, memories, and identity from a biological brain to a digital substrate, such as a computer or an artificial neural network. While the idea of mind transfer is often explored in science fiction, it raises profound philosophical, ethical, and scientific challenges, particularly concerning the problem of copying.

1. Mind Transfer and Identity: The idea of mind transfer raises fundamental questions about personal identity. If an individual's consciousness is transferred to a digital substrate, does the resulting digital entity retain the original person's identity, or does it create a separate, distinct identity? This philosophical "identity problem" explores whether the transferred mind is the same individual or merely a copy.

2. Problem of Copying: The problem of copying arises from the dilemma of creating a digital copy of a person's mind rather than transferring their consciousness. If a copy is made, two distinct entities coexist—the original biological person and the newly created digital version. From an outsider's perspective, both may seem identical, possessing the same memories and personality traits. However, from the subjective experience of each entity, their identities diverge.

3. Continuity of Consciousness: One of the central challenges in mind transfer is ensuring the continuity of consciousness. Consciousness is a continuous stream of experiences and sensations, creating a seamless sense of self. Transferring this continuity to a digital medium without interruption poses significant scientific and technical difficulties.

4. Preservation of Subjectivity: Preserving an individual's subjective experience during mind transfer is a critical concern. A faithful transfer should ensure that the subjective experience, sense of self, and emotional states remain intact during the process.

5. Ethical Considerations: Mind transfer raises ethical dilemmas regarding consent, autonomy, and the potential consequences for the transferred individual. Questions of whether a person's mind should be transferred without their explicit

consent or what rights and protections should be afforded to digital entities are complex and require careful consideration.

6. Technological Feasibility: The scientific and technological feasibility of mind transfer remains speculative and is subject to current limitations in neuroscience, computing power, and our understanding of consciousness.

7. Mind-Body Relationship: Mind transfer challenges conventional notions of the mind-body relationship. If a digital copy of a mind is created, it raises questions about the interconnectedness between consciousness and the biological brain.

In summary, mind transfer raises profound questions about personal identity, consciousness, and the nature of self. The problem of copying is a central philosophical challenge that calls into question the continuity of consciousness and subjective experience. While mind transfer remains speculative, the exploration of these concepts provides a platform for philosophical and ethical debates about the future of human identity and the potential implications of such technological advancements.

2.2. Neurotechnology and Cognitive Enhancement

Neurotechnology refers to the use of advanced technologies to interact with the nervous system and brain. Cognitive enhancement involves using neurotechnologies to improve cognitive functions, memory, attention, and learning. These emerging technologies hold promise for enhancing human capabilities, but ethical considerations regarding safety, equity, and potential risks need careful attention.

2.2.1. The Development of Neurotechnology and Its Impact on the Human Mind

The development of neurotechnology has significantly impacted the human mind, offering new possibilities for understanding brain functions and enhancing cognitive abilities. Techniques such as brain imaging (e.g., fMRI) provide insights into brain activity, facilitating research on cognition, emotions, and neurological disorders. Additionally, neurotechnologies like brain-computer interfaces (BCIs) enable direct communication between the brain and external devices, opening avenues for assistive technology and cognitive enhancement. While promising, ethical considerations, privacy concerns, and the responsible use of such technologies are essential to ensure their positive impact on human cognition and well-being.

2.2.1.1. Neural Interfaces and Neural Connections

Neural interfaces and neural connections are essential components of neurotechnology, enabling communication between the nervous system and external devices. These advancements have transformative implications for medical treatments, research, and cognitive enhancement.

1. Neural Interfaces: Neural interfaces, also known as brain-computer interfaces (BCIs), establish direct communication between the brain and external devices. BCIs can read neural signals and translate them into commands that control external devices or computer systems. There are different types of neural interfaces, including invasive (implanted directly into the brain), non-invasive (external devices that read brain signals), and partially invasive (implanted in the brain's surface).

2. Applications of Neural Interfaces:

Medical Treatments: Neural interfaces offer hope for people with motor disabilities, allowing them to control prosthetic limbs or assistive devices directly with their thoughts. They also hold potential for treating neurological conditions like Parkinson's disease and epilepsy.

Research and Neuroscience: Neural interfaces facilitate research on brain function and cognitive processes. They enable scientists to study brain activity during specific tasks, leading to a deeper understanding of the brain's complexity.

Cognitive Enhancement: Neural interfaces hold the potential for cognitive enhancement by connecting the brain to external technologies that augment memory, learning, and attention. However, ethical considerations about safety and consent are paramount in this context.

3. Neural Connections: Neural connections refer to the intricate networks of neurons in the brain that enable communication and information processing. The brain's vast neural connections underlie all cognitive processes, memory formation, and decision-making.

4. Neuroplasticity and Neural Connections: Neuroplasticity is the brain's ability to reorganize and form new neural connections in response to experiences, learning, and environmental changes. Neural interfaces and neurotechnologies can leverage neuroplasticity to adapt and integrate with the brain more effectively.

5. Ethical and Privacy Concerns: While neural interfaces offer promising opportunities, ethical concerns arise regarding consent, privacy, and potential misuse of such technologies. Safeguarding user autonomy, ensuring data security, and addressing the potential risks are critical considerations.

Neural interfaces and neural connections are crucial components of neurotechnology, revolutionizing medical treatments, cognitive research, and the potential for cognitive enhancement. As these technologies progress, striking a balance between technological advancement and ethical responsibility will be essential to maximize their benefits while ensuring the well-being and rights of individuals.

2.2.1.2. Enhancing Mental Abilities through Neurotechnology

Neurotechnology offers promising opportunities for enhancing various aspects of human mental abilities, revolutionizing the fields of medicine, cognitive science, and human-computer interaction. Here are some key ways in which neurotechnology is being explored for cognitive enhancement:

1. Memory Enhancement: Neurotechnologies like brain stimulation and neural implants hold potential for enhancing memory functions. Researchers are investigating the use of targeted brain stimulation to improve memory consolidation and retrieval. By understanding and modulating neural circuits involved in memory, we may unlock new ways to improve learning and recall.

2. Attention and Focus Improvement: Neurofeedback techniques allow individuals to monitor their brain activity and

learn to enhance attention and focus. By providing real-time feedback on brain states, individuals can train their minds to sustain attention on specific tasks, which could be valuable for various professions and educational settings.

3. Learning Acceleration: Neurofeedback and brain-computer interfaces (BCIs) hold promise for accelerating the learning process. These technologies can help optimize learning experiences, personalize educational content, and adapt learning approaches based on individual cognitive profiles.

4. Neuroplasticity and Skill Acquisition: Neuroplasticity, the brain's ability to rewire itself in response to experiences, is harnessed for skill acquisition. Neurotechnologies can stimulate specific brain regions to facilitate the development of motor skills, language acquisition, and other abilities.

5. Rehabilitation and Brain Repair: Neurotechnology plays a critical role in rehabilitation following brain injuries or strokes. Neural interfaces and brain stimulation therapies aid in restoring lost functions and promoting recovery.

6. Neurofeedback for Mental Health: Neurofeedback techniques show promise in managing mental health conditions such as anxiety, depression, and attention-deficit/hyperactivity disorder (ADHD). By providing individuals with real-time feedback

on their brain activity, they can learn to self-regulate and improve emotional well-being.

7. Brain-Computer Interfaces and Communication: For individuals with severe motor disabilities, brain-computer interfaces offer a means of communication and interaction with the world. These interfaces allow users to control external devices, communicate, and express themselves through their brain signals.

8. Ethical Considerations: As neurotechnology advances, ethical considerations are paramount. Questions about informed consent, data privacy, and the potential for unintended consequences must be addressed. Responsible development and transparent practices are essential to ensure the safe and ethical use of these technologies.

While neurotechnology shows immense promise for enhancing mental abilities, the responsible integration of these technologies is crucial. Combining scientific rigor, ethical guidelines, and user-centric design will pave the way for a future where cognitive enhancement technologies complement human potential, empower individuals, and improve overall well-being.

2.2.2. Improving and Enhancing Mental Abilities

Advancements in neurotechnology and cognitive science offer exciting opportunities for improving and enhancing various mental abilities. Techniques like brain stimulation, neural interfaces, and neurofeedback are being explored to enhance memory, attention, learning, and cognitive performance. These innovations hold promise for empowering individuals, revolutionizing education, and aiding medical rehabilitation. However, ethical considerations and responsible implementation are critical to ensure the safe and beneficial use of these technologies.

2.2.2.1. Mental Exercises and Neuroplasticity

Mental exercises and neuroplasticity are intertwined concepts that hold significant implications for enhancing cognitive abilities. Neuroplasticity refers to the brain's ability to reorganize and form new neural connections in response to learning, experiences, and environmental changes. Mental exercises leverage neuroplasticity to optimize brain function and enhance cognitive performance. Here are some key points about the relationship between mental exercises and neuroplasticity:

1. Harnessing Neuroplasticity: Neuroplasticity is a fundamental property of the brain that enables it to adapt and change throughout life. Neural pathways are strengthened with use and weakened with disuse, making the brain highly malleable and responsive to stimulation.

2. Benefits of Mental Exercises: Mental exercises involve targeted activities that challenge and stimulate specific cognitive

functions. Engaging in these exercises regularly can lead to improved memory, attention, problem-solving, and other cognitive skills.

3. Cognitive Training Programs: Cognitive training programs, often delivered through digital platforms, offer a structured approach to mental exercises. These programs are designed to target specific cognitive abilities and provide personalized training to enhance brain function.

4. Learning and Skill Acquisition: Learning new skills and acquiring knowledge induce neuroplastic changes in the brain. Whether it's learning a musical instrument, a new language, or solving complex puzzles, these activities stimulate neural growth and strengthen neural connections.

5. Adaptive Learning: Adaptive learning platforms use data and algorithms to personalize educational content based on individual learning patterns. By tailoring content to an individual's abilities and progress, adaptive learning optimizes the learning experience and capitalizes on neuroplasticity.

6. Neurofeedback and Mindfulness: Neurofeedback techniques, combined with mindfulness practices, enable individuals to become more aware of their brain activity and learn to self-regulate cognitive processes. These practices can enhance attention, reduce stress, and improve emotional well-being.

7. Lifelong Learning and Brain Health: Engaging in continuous learning and mental challenges throughout life helps maintain brain health and cognitive function. By regularly stimulating the brain, individuals can promote neuroplasticity and potentially reduce the risk of cognitive decline.

8. Ethical Considerations: While mental exercises and neuroplasticity offer promising benefits, it is essential to approach cognitive enhancement responsibly. Ensuring informed consent, data privacy, and avoiding false claims about cognitive improvement are crucial ethical considerations.

In conclusion, mental exercises and neuroplasticity are intertwined concepts that empower individuals to enhance cognitive abilities and optimize brain function. Leveraging the brain's inherent adaptability through mental exercises holds great potential for personal growth, lifelong learning, and improved cognitive performance across various stages of life.

2.2.2.2. Memory Enhancement and Learning Improvements

Memory enhancement and learning improvements are essential aspects of cognitive enhancement, and advancements in neurotechnology have opened new avenues for optimizing these cognitive functions. Here's a detailed exploration of memory enhancement and learning improvements through neurotechnological approaches:

1. Memory Enhancement: Neurotechnologies have demonstrated the potential to enhance various aspects of memory:

Brain Stimulation: Transcranial magnetic stimulation (TMS) and transcranial direct current stimulation (tDCS) are non-invasive brain stimulation techniques that have shown promise in boosting memory performance. These techniques modulate neural activity and promote synaptic plasticity, leading to improved memory consolidation and retrieval.

Neurofeedback Training: Neurofeedback training allows individuals to monitor their brain activity in real-time and learn to control specific brain regions associated with memory processes. Through neurofeedback, individuals can enhance memory functions by strengthening neural connections related to memory encoding and recall.

Pharmacological Interventions: Some pharmaceutical interventions have been explored to enhance memory consolidation and retrieval. Certain drugs and compounds target neurotransmitter systems involved in memory processes, potentially leading to memory improvement.

2. Learning Improvements: Neurotechnologies offer exciting opportunities for optimizing learning processes:

Adaptive Learning Platforms: Adaptive learning systems utilize data analytics and artificial intelligence to personalize educational content based on an individual's learning patterns. These platforms identify areas of strength and weakness and provide tailored learning experiences, maximizing learning efficiency.

Virtual Reality (VR) Learning: VR-based learning experiences provide immersive and interactive environments for learners. By engaging multiple senses, VR learning enhances information retention and understanding, particularly for complex subjects.

Cognitive Training Programs: Cognitive training programs, delivered through digital applications, target specific cognitive skills such as attention, problem-solving, and spatial reasoning. Engaging in these programs regularly can lead to more efficient and effective learning outcomes.

Multimodal Learning: Combining various learning modalities, such as visual, auditory, and kinesthetic approaches, can optimize information processing and memory encoding. Utilizing multiple sensory channels enhances learning retention and comprehension.

3. Educational Applications: Neurotechnological approaches for memory enhancement and learning improvements have broad applications in education:

Enhanced Educational Curricula: Integrating neuroscientific principles into educational curricula can optimize learning environments, promoting better memory retention and knowledge acquisition.

Personalized Learning Plans: Neurotechnological insights can inform personalized learning plans for students, considering individual cognitive strengths and weaknesses. Tailored approaches maximize learning potential for each student.

Lifelong Learning Initiatives: Encouraging lifelong learning through neurotechnological interventions can support cognitive health and improve cognitive flexibility across all ages.

4. Ethical Considerations: As neurotechnological interventions for memory enhancement and learning improvements advance, ethical considerations are paramount. Ensuring informed consent, data privacy, and equitable access to cognitive enhancements are essential to fostering responsible use and minimizing potential risks.

In summary, memory enhancement and learning improvements through neurotechnologies offer transformative possibilities for optimizing cognitive functions and educational outcomes. Ethical and responsible implementation of these interventions will play a critical role in harnessing their benefits and promoting positive impacts on individual learning and cognitive well-being.

2.2.3. The Application of Neurotechnology in Education and Learning

Neurotechnology holds significant potential for transforming education and learning experiences. By leveraging brain-computer interfaces, adaptive learning platforms, and neurofeedback techniques, neurotechnology enhances personalized learning, optimizes educational content, and promotes cognitive development. These applications empower students with tailored learning experiences and open new horizons for optimizing educational outcomes.

2.2.3.1. Neurotechnology-Supported Education and Teaching Methods

Neurotechnology-supported education and teaching methods integrate advancements in cognitive science and brain-computer interfaces to enhance the learning experience. These innovative approaches leverage neuroscientific insights to optimize educational practices and improve student outcomes. Here are some key areas where neurotechnology is transforming education and teaching methods:

1. Adaptive Learning Platforms: Adaptive learning platforms utilize data analytics and machine learning algorithms to personalize educational content for each student. By continuously assessing an individual's performance and learning preferences, these platforms adapt the curriculum to meet the student's specific needs, strengths, and weaknesses. This personalized approach enhances engagement and knowledge retention, ensuring that students progress at their optimal pace.

2. Neurofeedback Training: Neurofeedback techniques enable students to become aware of their brain activity and learn to self-regulate cognitive processes. This technology provides real-time feedback on brain states associated with attention, focus, and emotional regulation. Integrating neurofeedback training in the classroom helps students develop better self-awareness and self-control, leading to improved learning and emotional well-being.

3. Virtual Reality (VR) Learning: Virtual reality (VR) technology offers immersive and interactive learning experiences that engage multiple senses. By creating realistic virtual environments, VR learning enhances students' understanding and retention of complex concepts. It allows students to explore historical events, travel to distant places, and participate in hands-on simulations, making learning more exciting and impactful.

4. Brain-Based Teaching Strategies: Brain-based teaching strategies align with neuroscientific principles to optimize how students learn. These strategies incorporate movement, visualization, storytelling, and multisensory approaches to enhance memory encoding and retrieval. By catering to diverse learning styles, brain-based teaching fosters better understanding and long-term retention of information.

5. Cognitive Training Programs: Cognitive training programs target specific cognitive skills, such as attention, problem-solving, and memory. These programs engage students in brain exercises and puzzles that stimulate neural pathways associated with cognitive functions. Integrating cognitive training into the curriculum enhances critical thinking abilities and cognitive flexibility.

6. Mindfulness Practices: Introducing mindfulness practices in the classroom promotes students' emotional well-being and resilience. Mindfulness techniques, such as meditation and breathing exercises, help reduce stress and improve concentration, creating a conducive learning environment.

7. Brain-Computer Interfaces for Communication: Brain-computer interfaces (BCIs) enable communication for students with severe motor disabilities. BCIs allow these students to control computers or communication devices directly with their brain

signals, facilitating active participation in the learning process and promoting inclusivity.

8. Neuroscience-Informed Curriculum Design: Neuroscience insights inform curriculum design, ensuring that educational content aligns with how the brain processes information. By structuring lessons based on the brain's natural learning processes, educators optimize knowledge acquisition and retention.

The application of neurotechnology in education enhances teaching methods, promotes personalized learning experiences, and fosters student engagement. As neuroscientific research advances, the integration of ethical and evidence-based neurotechnology-supported education will continue to shape the future of learning, making education more effective, inclusive, and enriching for all students.

2.3. Brain-Computer Interfaces and Digital Integration

Brain-computer interfaces (BCIs) are revolutionary technologies that establish direct communication between the human brain and external devices or digital systems. BCIs enable individuals to control technology using their thoughts and brain signals. This digital integration opens up new possibilities for medical applications, communication for individuals with disabilities, and novel ways of interacting with computers and virtual environments. However, ethical considerations, data privacy, and responsible use are crucial aspects to address as BCIs become more prevalent in our digital world.

2.3.1. Principles and Technologies of Brain-Computer Interfaces

Brain-computer interfaces (BCIs) operate on the principles of capturing, interpreting, and translating brain signals into actionable commands for external devices. These interfaces use various technologies to achieve seamless communication between the brain and digital systems. Electroencephalography (EEG), invasive brain implants, and non-invasive brain stimulation are among the key technologies employed in BCIs. These innovations pave the way for groundbreaking applications in medical treatments, assistive technology, and human-computer interaction.

2.3.1.1. Brain-Computer Interfaces and EEG Technology

Brain-computer interfaces (BCIs) leverage EEG (electroencephalography) technology to establish a direct communication pathway between the brain and external devices. EEG is a non-invasive method for recording electrical activity in the brain through electrodes placed on the scalp. Here's a detailed exploration of BCIs and EEG technology:

1. EEG and Brain Signal Recording: EEG measures the electrical activity generated by the brain's neurons. When neurons communicate, they produce electrical impulses that can be detected by electrodes on the scalp. EEG records these brain signals in the form of brainwaves, which represent different brain states and cognitive activities.

2. Non-Invasive Nature of EEG: One of the significant advantages of EEG is its non-invasiveness. Unlike invasive techniques that require implants within the brain, EEG electrodes are placed on the scalp, making it a safer and more accessible option for brain signal recording.

3. Real-Time Brain Signal Analysis: EEG allows for real-time monitoring and analysis of brain activity. The recorded brainwaves can be processed instantly to detect specific patterns, such as attention levels, motor intentions, or emotional states.

4. BCI Applications using EEG: BCIs that utilize EEG technology have diverse applications:

Assistive Technology: EEG-based BCIs enable individuals with severe motor disabilities to control computers, communication devices, or robotic prosthetics using their brain signals.

Medical Treatments: EEG-based BCIs aid in neurofeedback therapy for conditions like epilepsy, ADHD, and other neurological disorders, promoting self-regulation and cognitive improvement.

Cognitive Enhancement: EEG neurofeedback training allows individuals to enhance attention, memory, and emotional regulation through self-regulated brain activity.

Human-Computer Interaction: EEG-based BCIs can enable hands-free control of virtual reality environments, video games, or other digital interfaces, enhancing the user experience.

5. Challenges and Future Developments: Despite its advantages, EEG-based BCIs face challenges such as signal noise, limited spatial resolution, and the need for calibration. Research is ongoing to improve the accuracy, reliability, and speed of EEG-based BCIs. Additionally, combining EEG with other neuroimaging techniques, such as fNIRS (functional near-infrared spectroscopy) or fMRI (functional magnetic resonance imaging), may lead to more comprehensive brain-computer communication systems.

6. Ethical Considerations: Ethical considerations surrounding EEG-based BCIs include issues of data privacy, consent, and potential misuse of brain data. Ensuring the confidentiality and security of sensitive brain signals is crucial for responsible BCI development and use.

EEG technology plays a pivotal role in brain-computer interfaces, offering a non-invasive method for recording brain signals and enabling various applications in medical treatments, assistive technology, and human-computer interaction. Ongoing research and ethical considerations will drive further advancements in EEG-based BCIs, shaping the future of brain-computer communication and its impact on human well-being.

2.3.1.2. Digital Communication and Brain-Skin Interfaces

Brain-skin interfaces represent an emerging area of research that aims to establish communication between the brain and external digital devices through the skin. These interfaces use various technologies to enable bidirectional data transmission, offering promising applications in both medical and non-medical domains. Here's a detailed exploration of brain-skin interfaces and their potential for digital communication:

1. Brain-Skin Interface Technology: Brain-skin interfaces involve the use of wearable devices or implants that interact with the brain and skin to facilitate communication. These interfaces can record brain signals and deliver sensory feedback through the skin, allowing for seamless two-way information transfer.

2. Bidirectional Communication: The bidirectional nature of brain-skin interfaces enables not only the recording of brain signals but also the delivery of sensory feedback to the user. This feedback can be in the form of tactile sensations, vibrations, or other haptic feedback, enhancing the interactive experience.

3. Non-Invasive and Minimally Invasive Approaches: Brain-skin interfaces can be designed to be non-invasive, involving wearable devices that sit on the skin's surface. Alternatively, they can be minimally invasive, with thin and flexible electrodes implanted under the skin to establish a more direct connection with the nervous system.

4. Applications in Medicine: Brain-skin interfaces hold promising applications in the medical field:

Prosthetics and Assistive Technology: Brain-skin interfaces can enhance the control and sensory feedback of prosthetic limbs, allowing users to manipulate and sense objects more naturally.

Neurorehabilitation: Brain-skin interfaces can aid in neurorehabilitation by providing sensory feedback to patients with neurological disorders, promoting motor learning and recovery.

Pain Management: Brain-skin interfaces may be used for non-pharmacological pain management through neuromodulation techniques delivered via the skin.

5. Non-Medical Applications: Beyond medical applications, brain-skin interfaces have potential non-medical uses:

Virtual and Augmented Reality: Brain-skin interfaces can enhance immersion in virtual and augmented reality experiences by providing haptic feedback, creating a more realistic and interactive environment.

Gaming and Entertainment: Brain-skin interfaces could revolutionize gaming and entertainment by enabling players to experience tactile feedback and sensations related to in-game events.

6. Challenges and Future Prospects: Brain-skin interfaces are still in their early stages of development, and several challenges need to be addressed. These include ensuring the accuracy and reliability of signal transmission, minimizing invasiveness for long-term use, and optimizing the haptic feedback for a natural and intuitive experience.

7. Ethical Considerations: As brain-skin interfaces develop, ethical considerations concerning data privacy, informed consent, and potential long-term effects on the nervous system must be carefully addressed to safeguard user well-being.

Brain-skin interfaces represent a cutting-edge field with promising potential for digital communication between the brain and external devices. These interfaces offer exciting applications in both medical and non-medical domains, paving the way for innovative solutions in prosthetics, neurorehabilitation, virtual reality, and beyond. As research progresses, addressing challenges and ethical considerations will be key to unlocking the full potential of brain-skin interfaces in enhancing human-computer interaction and improving quality of life.

2.3.2. Mind Control in Virtual and Augmented Reality

Mind control in virtual and augmented reality refers to the ability to interact with these immersive environments using brain-computer interfaces (BCIs) or other neurotechnologies. BCIs allow users to control virtual objects, navigate virtual worlds, and interact with digital interfaces through their brain signals. This technology offers new dimensions of user engagement and a more intuitive way to experience and manipulate virtual and augmented reality environments. However, ethical considerations, user consent, and data privacy are essential aspects to address as mind control technology advances in these virtual realms.

2.3.2.1. Mind-Controlled Interactions in Virtual Worlds

Mind-controlled interactions in virtual worlds represent a groundbreaking advancement in human-computer interaction, enabling users to manipulate virtual environments and objects using their brain signals. Leveraging brain-computer interfaces (BCIs), this technology allows for intuitive and immersive experiences in virtual reality (VR) environments. Here's a detailed exploration of mind-controlled interactions in virtual worlds:

1. Brain-Computer Interfaces (BCIs) in VR: BCIs used in VR environments are non-invasive or minimally invasive devices that record and interpret brain signals. These signals are then translated into actionable commands for the virtual world. Non-invasive BCIs use EEG technology, while minimally invasive BCIs may involve implanted electrodes for a more direct brain signal connection.

2. Telekinesis and Object Manipulation: Mind-controlled interactions in VR enable users to experience a form of telekinesis, where they can manipulate virtual objects through their thoughts. For example, users can pick up, move, and interact with objects in the virtual environment solely using their brain signals, creating a more immersive and intuitive experience.

3. Avatar Control and Body Movements: Mind-controlled interactions extend to avatar control within VR. Users can control

the movements of their virtual avatars using their brain signals, allowing for a more natural embodiment and movement within the virtual space.

4. Immersive Experiences and Presence: By enabling mind-controlled interactions, VR experiences become more immersive and provide a sense of presence, where users feel as if they are truly present within the virtual world. This enhanced sense of immersion enhances the overall user experience.

5. Enhanced Accessibility and Inclusivity: Mind-controlled interactions in VR hold great promise for individuals with physical disabilities, as it provides an alternative and more accessible means of interaction. BCIs offer opportunities for people with motor impairments to navigate and interact with virtual worlds independently.

6. Ethical Considerations: While mind-controlled interactions in virtual worlds offer exciting possibilities, ethical considerations are crucial. Consent, data privacy, and ensuring the technology's safety and reliability are paramount to protect user well-being and rights.

7. Research and Advancements: Ongoing research is focused on improving the accuracy and responsiveness of mind-controlled interactions in VR. Advancements in BCI technology and

neural decoding algorithms continue to push the boundaries of what is possible within virtual worlds.

In conclusion, mind-controlled interactions in virtual worlds revolutionize human-computer interaction, offering a more natural and intuitive way to navigate and manipulate virtual environments. This technology enhances immersion, accessibility, and the overall VR experience. As research progresses and ethical guidelines are established, mind-controlled interactions in virtual worlds will undoubtedly shape the future of virtual reality and open new possibilities for human-computer interaction.

2.3.2.2. Augmented Reality and Brain Commands

Augmented reality (AR) combined with brain commands represents a cutting-edge synergy of technology, enabling users to interact with digital overlays in the real world using their brain signals. This innovative integration leverages brain-computer interfaces (BCIs) to provide a seamless and intuitive AR experience. Here's a detailed exploration of augmented reality and brain commands:

1. Brain-Computer Interfaces (BCIs) in Augmented Reality: BCIs used in augmented reality capture and interpret brain signals to facilitate interactions with digital content superimposed onto the physical world. These interfaces can be non-invasive, utilizing

EEG technology, or minimally invasive, involving implanted electrodes for more direct neural connections.

2. Hands-Free Interactions: One of the key advantages of augmented reality with brain commands is hands-free interaction. Users can navigate, manipulate, and interact with AR elements using their brain signals, freeing their hands for other tasks.

3. Spatial Awareness and Contextual Interaction: AR with brain commands allows for spatially aware and contextual interactions. Users can simply look at an AR element or think about a specific command to trigger actions, enhancing the user experience and making interactions more seamless.

4. Virtual Object Interaction: Users can interact with virtual objects in the real world through brain commands. For example, users can move, resize, rotate, and manipulate virtual 3D models using their thoughts, creating a more immersive and interactive AR experience.

5. Information Retrieval and Overlay: AR with brain commands enables users to retrieve information and digital overlays in real-time. By simply thinking about specific commands or objects, users can access relevant digital information superimposed onto their physical environment.

6. Personalized AR Experiences: BCIs enable personalized AR experiences based on user preferences and cognitive states. AR content can be tailored to individual interests, making the interaction with digital overlays more relevant and engaging.

7. Ethical Considerations: Ethical considerations surrounding AR and brain commands include privacy, consent, and data security. Ensuring that users have control over their brain data and that it is used responsibly is crucial in the development and implementation of this technology.

8. Future Prospects: As BCI technology and AR applications advance, the integration of brain commands in AR will likely become more seamless and prevalent. Continued research and development will drive the technology's accessibility, safety, and efficiency.

Augmented reality combined with brain commands opens up a new frontier of human-computer interaction, offering hands-free and intuitive ways to interact with digital content in the real world. This technology has promising applications in various domains, including information retrieval, virtual object manipulation, and personalized AR experiences. Responsible development and ethical considerations are essential to maximize the potential benefits and ensure user well-being in this exciting intersection of augmented reality and brain commands.

2.3.3. Brain Implants and the Ethical Challenges of Brain-Computer Integration

Brain implants are advanced technologies that establish direct communication between the brain and external devices, facilitating brain-computer integration. While brain implants hold immense potential for medical treatments and assistive technology, they also raise significant ethical challenges. Concerns include data privacy, informed consent, potential risks of invasive procedures, and ensuring equitable access to these technologies. Striking a balance between scientific progress and responsible use is crucial to address these ethical challenges and maximize the benefits of brain-computer integration.

2.3.3.1. Brain Implants and the Neural Modifiability

Brain implants have the remarkable ability to modify neural activity and connectivity, enabling direct communication between the brain and external devices. These devices, known as neural implants or brain-computer interfaces (BCIs), hold great promise for medical treatments and assistive technologies. Here's a detailed exploration of brain implants and their impact on neural modifiability:

1. Neural Modifiability and Brain-Computer Integration: Brain implants interface with the brain's neural circuits, allowing for bidirectional communication. They can both record neural

activity and stimulate specific brain regions to influence brain function. This neural modifiability is central to brain-computer integration and the potential to enhance cognitive abilities and motor control.

2. Medical Applications and Neuroprosthetics: Brain implants have transformative applications in medical treatments. Neuroprosthetics, such as brain-controlled robotic limbs, help restore movement and independence to individuals with severe motor disabilities. Neural implants also hold promise in treating neurological disorders, epilepsy, and chronic pain through targeted neuromodulation.

3. Cognitive Enhancement and Neurofeedback: Neural implants offer possibilities for cognitive enhancement through neurofeedback training. By providing real-time information about brain activity, individuals can learn to self-regulate neural patterns associated with attention, memory, and emotional states, potentially improving cognitive performance.

4. Ethical Considerations: The neural modifiability enabled by brain implants raises significant ethical concerns. Ensuring informed consent, privacy of neural data, and equitable access to these technologies is essential. The potential for unintended consequences or misuse of neural modifiability requires thorough consideration and responsible governance.

5. Long-Term Effects and Risks: Long-term effects of brain implants on neural circuits and overall brain health require extensive study. The risk of infections, immune responses, or unintended side effects from chronic implantation must be carefully assessed and mitigated.

6. Brain Plasticity and Adaptation: Brain implants can influence neural plasticity, the brain's ability to reorganize and adapt in response to experiences and stimuli. Understanding how neural circuits adapt to brain implants is crucial for optimizing long-term outcomes and maximizing benefits.

7. Public Perception and Acceptance: The acceptance and adoption of brain implants in society may be influenced by public perception, cultural attitudes, and stigmatization. Ethical discussions and public engagement are necessary to foster understanding and acceptance of brain-computer integration technologies.

8. Responsible Innovation: Responsible innovation is essential in the development and deployment of brain implants. Robust regulatory frameworks, interdisciplinary research, and ongoing ethical assessments are crucial to guide the responsible use of neural modifiability.

Brain implants and neural modifiability present exciting opportunities for medical advancements and cognitive enhancement. However, addressing ethical considerations, understanding the long-term effects, and promoting responsible innovation are essential to harness the potential of brain-computer integration while safeguarding individual well-being and societal values.

2.3.3.2. Personal Privacy and Social Control in Brain-Computer Integration

Brain-computer integration technologies, such as brain implants and brain-computer interfaces (BCIs), raise critical concerns about personal privacy and social control. These technologies have the potential to access, interpret, and manipulate individuals' neural data, presenting ethical challenges related to data privacy, autonomy, and individual agency. Here's a detailed exploration of the issues surrounding personal privacy and social control in brain-computer integration:

1. Invasive Nature and Data Privacy: Brain implants and invasive BCIs involve direct access to the brain's neural signals. The sensitive and personal nature of neural data raises concerns about data privacy, as unauthorized access or misuse of this information could have serious consequences for individuals' privacy and security.

2. Informed Consent and Autonomy: Obtaining informed consent is essential for brain-computer integration procedures.

Individuals should have a comprehensive understanding of the potential risks, benefits, and implications of using these technologies. Preserving autonomy allows individuals to make informed decisions about the use and sharing of their neural data.

3. Neurosurveillance and Social Control: The capability to monitor and analyze neural data may lead to neurosurveillance, where individuals' thoughts and emotions could be observed and interpreted. This raises concerns about potential social control, as powerful entities might seek to manipulate or influence people's behaviors based on their neural data.

4. Data Ownership and Control: Determining data ownership and control becomes complex with brain-computer integration. Clear guidelines must be established to ensure that individuals retain ownership and control over their neural data, protecting them from undue external influence or manipulation.

5. Brain Data Security: Brain data security is paramount to prevent unauthorized access or hacking of neural information. Ensuring robust encryption and secure storage of neural data is crucial to safeguard individuals' privacy.

6. Discrimination and Stigmatization: The use of brain-computer integration data may lead to discrimination and stigmatization based on individuals' neurological characteristics.

Protecting against such biases is essential to promote a fair and equitable society.

7. Regulatory Framework and Governance: Developing comprehensive regulatory frameworks and governance mechanisms is crucial to address personal privacy and social control concerns. Governments and international bodies must collaborate to establish ethical guidelines and standards for brain-computer integration technologies.

8. Public Awareness and Education: Promoting public awareness and education about brain-computer integration technologies is vital to foster understanding and informed discussions about the implications of these advancements on personal privacy and social dynamics.

Brain-computer integration technologies have the potential to significantly impact personal privacy and social control. Ethical considerations, data privacy protections, informed consent, and responsible governance are critical in navigating these complex issues. Striking a balance between the benefits of brain-computer integration and safeguarding individual rights and autonomy is essential to ensure a positive and equitable impact on society.

CHAPTER 3

Biological Enhancement and Longevity

3.1. The Aging Process and Its Biological Causes

Aging is a natural process involving the gradual decline of physiological functions and increased vulnerability to diseases. Cellular senescence, telomere shortening, genetic factors, oxidative stress, inflammation, mitochondrial dysfunction, and epigenetic changes contribute to aging. Understanding these biological causes helps develop strategies for healthy aging.

3.1.1. Aging and the Biological Mechanisms of Aging

Aging is a complex process influenced by various biological mechanisms. Over time, cells undergo senescence, telomeres shorten, and genetic factors impact susceptibility to age-related conditions. Oxidative stress, inflammation, mitochondrial dysfunction, and epigenetic changes play significant roles in the aging process. Understanding these mechanisms is vital for developing interventions to promote healthy aging and address age-related challenges.

3.1.1.1. Cellular Damages and DNA Breaks

Cellular damages and DNA breaks are fundamental contributors to the aging process. As cells replicate and face environmental stressors throughout life, various forms of damage accumulate, leading to age-related changes. Here's a detailed exploration of cellular damages and DNA breaks in the context of aging:

1. Replicative Senescence and Telomere Shortening: During each cell division, telomeres, protective caps at the ends of chromosomes, experience slight shortening. Over time, this shortening accumulates, leading to replicative senescence—a state where cells stop dividing or functioning optimally. This process limits tissue regeneration and contributes to aging.

2. Oxidative Stress and Free Radicals: Cells generate energy through metabolic processes, producing free radicals as byproducts. These highly reactive molecules can damage cellular structures, including DNA. Over time, accumulated oxidative stress results in DNA breaks and contributes to cellular dysfunction.

3. DNA Damage Response and Repair Mechanisms: Cells have intricate DNA repair mechanisms to address damages caused by internal and external factors. However, as aging progresses, the efficiency of these repair processes declines, leading to the persistence of DNA breaks and damage.

4. Accumulation of Cellular Waste Products: Throughout life, cells generate waste products that need to be cleared for proper function. Over time, the accumulation of waste products,

such as lipofuscin, impairs cellular function and contributes to aging.

5. Inflammatory Responses and Cellular Senescence: Chronic inflammation can induce cellular senescence—a state where cells stop dividing and undergo functional changes. These senescent cells release pro-inflammatory molecules, further promoting tissue dysfunction and aging.

6. Mitochondrial Dysfunction and DNA Damage: Mitochondria, the cell's powerhouses, can generate reactive oxygen species during energy production. These species contribute to oxidative stress and DNA damage, impacting mitochondrial function and cellular health.

7. Environmental Factors and DNA Breaks: Exposure to environmental toxins, radiation, and other external factors can induce DNA breaks and contribute to cellular damage over time.

Understanding the mechanisms of cellular damages and DNA breaks in aging is crucial for developing strategies to mitigate age-related changes and promote healthier aging. Targeting these processes may offer potential avenues for interventions and therapies to address age-related conditions and extend the quality of life in older individuals.

3.1.1.2. Oxidative Stress and Antioxidant Defense

Oxidative stress is a significant biological mechanism contributing to the aging process. It occurs when there is an imbalance between the production of reactive oxygen species (ROS) and the body's ability to neutralize or repair the resulting damage. Here's a detailed exploration of oxidative stress and the antioxidant defense system:

1. Reactive Oxygen Species (ROS): ROS are chemically reactive molecules containing oxygen that are generated as natural byproducts of cellular metabolism. While they play essential roles in cell signaling and immune responses, excessive ROS production can lead to cellular damage and aging.

2. Sources of ROS: ROS can be produced through various cellular processes, including mitochondrial respiration, inflammation, and environmental exposures (e.g., pollution, UV radiation, smoking).

3. Cellular Damage by ROS: ROS can cause oxidative damage to cellular components, including lipids, proteins, and DNA. This damage can disrupt cellular function and contribute to age-related changes and disease development.

4. Antioxidant Defense System: The body has a sophisticated antioxidant defense system to counteract the damaging effects of ROS. This system includes enzymes, such as

superoxide dismutase (SOD), catalase, and glutathione peroxidase, as well as non-enzymatic antioxidants like vitamins C and E.

5. Enzymatic Antioxidants: Enzymatic antioxidants, such as SOD, catalase, and glutathione peroxidase, convert ROS into less harmful molecules, neutralizing their damaging potential.

6. Non-enzymatic Antioxidants: Non-enzymatic antioxidants, like vitamins C and E, directly scavenge ROS, preventing them from causing cellular damage.

7. Cellular Defense and Repair: The antioxidant defense system also plays a role in cellular repair processes. For example, damaged DNA is repaired by specific enzymes, reducing the risk of mutations and cellular dysfunction.

8. Impairment of Antioxidant Defense in Aging: With aging, the antioxidant defense system may become less efficient, leading to a decreased ability to neutralize ROS effectively. This imbalance can contribute to increased oxidative damage and accelerated aging.

9. Lifestyle Factors and Antioxidant Balance: Lifestyle factors, such as diet, exercise, and stress levels, can influence the body's antioxidant balance. A diet rich in antioxidants, regular physical activity, and stress management can support the antioxidant defense system and promote healthier aging.

Understanding the role of oxidative stress and the antioxidant defense system is crucial for developing interventions and lifestyle strategies to mitigate age-related cellular damage and promote overall well-being in aging individuals. Targeting oxidative stress through various means, including antioxidant-rich diets and lifestyle modifications, may offer potential avenues to improve the quality of life during aging.

3.1.2. The Influence of Genetic and Environmental Factors on Aging

Aging is influenced by a combination of genetic and environmental factors. Genetic factors play a role in an individual's susceptibility to age-related conditions, the rate of cellular aging, and the efficiency of repair mechanisms. Environmental factors, such as lifestyle choices, diet, exposure to pollutants, and stress levels, also impact the aging process. The interplay between genetics and the environment determines how individuals age and experience age-related changes. Understanding these influences is essential for developing personalized approaches to promote healthy aging and address age-related challenges

3.1.2.1. Genetic Predispositions and Aging

Genetic predispositions play a crucial role in the aging process, influencing an individual's susceptibility to age-related conditions and the rate of cellular aging. Here's a detailed exploration of how genetic factors contribute to aging:

1. Genes and Cellular Senescence: Certain genes regulate the process of cellular senescence, where cells stop dividing and undergo functional changes. Genetic variations can influence the onset of cellular senescence, impacting tissue repair and regeneration during aging.

2. Telomeres and Genetic Stability: Genes also play a role in maintaining telomere length, which protects chromosomes from deterioration during cell division. Shortened telomeres, resulting from genetic factors, can accelerate cellular aging and contribute to age-related conditions.

3. DNA Repair Genes and Cellular Damage: Genetic variations in DNA repair genes can affect the efficiency of DNA damage repair mechanisms. Reduced repair capacity leads to the accumulation of DNA damage over time, contributing to cellular dysfunction and aging.

4. Susceptibility to Age-Related Diseases: Certain genetic variants can increase an individual's susceptibility to age-related diseases, such as cardiovascular disorders, neurodegenerative conditions, and cancer.

5. Inflammation and Immune Response Genes: Genetic factors influence the body's inflammatory and immune responses. Chronic inflammation, driven by genetic predispositions, can contribute to age-related tissue damage and diseases.

6. Longevity-Associated Genes: Researchers have identified longevity-associated genes, also known as "longevity genes," that appear to confer a protective effect against age-related conditions, promoting healthy aging.

7. Familial Longevity Clusters: Some families exhibit a clustering of exceptional longevity, suggesting a strong genetic component in longevity. Studying these families can provide insights into the genetic basis of healthy aging.

8. Gene-Environment Interactions: It is important to note that genetic predispositions interact with environmental factors in shaping the aging process. Lifestyle choices, diet, physical activity, and exposure to environmental stressors can modify the expression of certain genes and influence how individuals age.

Understanding the genetic predispositions to aging offers valuable insights into the underlying mechanisms of age-related changes. As genetic research advances, personalized approaches to healthy aging may be developed, considering an individual's unique genetic profile and lifestyle factors. It is essential to integrate genetic information with environmental considerations to promote healthier aging and address age-related challenges effectively.

3.1.2.2. Environmental Stress and Accelerated Aging

Environmental stressors can accelerate the aging process by contributing to cellular damage, promoting inflammation, and affecting overall health. Here's a detailed exploration of how environmental factors influence aging:

1. Oxidative Stress and Pollutant Exposure: Exposure to environmental pollutants, such as air pollutants, heavy metals, and pesticides, can generate oxidative stress in the body. Oxidative stress leads to cellular damage and accelerates aging processes.

2. Lifestyle Choices and Aging: Unhealthy lifestyle choices, such as a sedentary lifestyle, poor diet, smoking, and excessive alcohol consumption, can negatively impact health and accelerate aging. These choices contribute to chronic inflammation and increase the risk of age-related diseases.

3. Sun Exposure and Skin Aging: Excessive exposure to ultraviolet (UV) radiation from the sun can lead to premature skin aging, including wrinkles, sunspots, and loss of skin elasticity.

4. Stress and Cellular Aging: Chronic stress can accelerate cellular aging by shortening telomeres and increasing oxidative

stress. Prolonged stress responses may contribute to age-related diseases and overall health decline.

5. Diet and Cellular Health: A diet lacking essential nutrients and antioxidants can compromise cellular health and contribute to age-related damage. Conversely, a balanced and nutrient-rich diet supports cellular function and promotes healthy aging.

6. Physical Activity and Aging: Regular physical activity is associated with healthier aging outcomes. Exercise can improve cardiovascular health, enhance mitochondrial function, and reduce inflammation, all of which contribute to better aging.

7. Sleep and Restoration: Poor sleep quality and insufficient sleep duration can impair cellular repair processes and disrupt the body's ability to restore and rejuvenate, potentially accelerating aging.

8. Psychological Factors and Aging: Psychological well-being, social connections, and positive mental states have been linked to healthier aging outcomes. Supportive social networks and emotional resilience may buffer the effects of stress on aging.

9. Epigenetics and Environmental Influence: Environmental factors can influence epigenetic modifications, affecting gene expression without altering the DNA sequence. Epigenetic changes

may play a role in the accelerated aging observed in response to environmental stressors.

10. Cumulative Effects: The cumulative effects of environmental stressors over time can lead to accelerated aging and increased vulnerability to age-related diseases.

Understanding the impact of environmental stress on aging is essential for promoting healthy aging and developing interventions to counteract age-related challenges. Adopting a healthy lifestyle, managing stress, and minimizing exposure to harmful environmental factors are essential strategies to support optimal aging and overall well-being.

3.1.3. The Relationship Between Aging and Diseases and Their Treatment

Aging is closely linked to the development of various age-related diseases. As individuals age, their risk of conditions such as cardiovascular disease, neurodegenerative disorders, cancer, and diabetes increases. The aging process can exacerbate the progression of these diseases. However, advancements in medical treatments and interventions have improved disease management and quality of life for aging individuals. Understanding the complex relationship between aging and diseases is crucial for developing targeted therapies and personalized approaches to address age-related health challenges effectively.

3.1.3.1. The Connection between Aging and Diseases

Aging and diseases have a complex and intertwined relationship. As individuals age, they become more susceptible to various age-related diseases, and the presence of these diseases can accelerate the aging process. Here's a detailed exploration of the connection between aging and diseases:

1. Increased Disease Risk with Age: As individuals grow older, their risk of developing age-related diseases significantly increases. These diseases include cardiovascular conditions, neurodegenerative disorders (e.g., Alzheimer's and Parkinson's), cancer, diabetes, osteoporosis, and others.

2. Cellular Aging and Disease Development: Cellular aging, including cellular damage, telomere shortening, and impaired DNA repair mechanisms, contributes to the development and progression of age-related diseases. Accumulated cellular damage can impair organ function and increase disease risk.

3. Inflammation and Age-Related Diseases: Chronic inflammation, which becomes more prevalent with aging, plays a crucial role in the development of age-related diseases. Inflammatory processes can damage tissues and organs, contributing to various health conditions.

4. Shared Risk Factors: Some age-related diseases share common risk factors. For example, obesity, sedentary lifestyles,

and poor dietary habits can increase the risk of both cardiovascular disease and diabetes.

5. Disease Impact on Aging: Age-related diseases can further accelerate the aging process. The burden of chronic diseases can lead to physical and cognitive decline, reduce life expectancy, and diminish overall quality of life.

6. Treatments and Disease Management: Medical advancements have improved disease management and treatment, allowing individuals with age-related diseases to live longer and healthier lives. Effective disease management can mitigate the impact of these conditions on aging.

7. Holistic Approaches to Aging and Disease: Holistic approaches that address lifestyle factors, such as diet, exercise, and stress management, can influence both aging and disease development. A healthy lifestyle can promote successful aging and reduce the risk of age-related diseases.

8. Personalized Medicine and Aging: Understanding the individual variations in disease development and response to treatments is essential. Personalized medicine approaches consider an individual's unique genetic makeup and lifestyle factors to tailor treatments and interventions.

Understanding the connection between aging and diseases is vital for promoting healthy aging and improving disease management in older individuals. Integrating preventive measures, early detection, and targeted therapies can significantly impact the trajectory of age-related diseases and enhance overall well-being during aging.

3.1.3.2. Anti-Aging Treatments and Disease Prevention

Anti-aging treatments and disease prevention strategies play crucial roles in promoting healthy aging and reducing the burden of age-related diseases. These interventions aim to slow down the aging process, enhance quality of life, and reduce the risk of developing age-associated health conditions. Here's a detailed exploration of anti-aging treatments and disease prevention:

1. Lifestyle Modifications: Healthy lifestyle choices, such as regular physical activity, a balanced diet rich in antioxidants and nutrients, stress management, and adequate sleep, can positively influence the aging process and reduce the risk of age-related diseases.

2. Antioxidant Supplements: Antioxidant supplements, such as vitamins C and E, may help neutralize harmful free radicals and reduce oxidative stress, potentially slowing down cellular aging.

3. Hormone Replacement Therapies: Hormone replacement therapies, such as estrogen replacement in postmenopausal

women, can help alleviate age-related symptoms and improve overall health. However, these treatments should be carefully evaluated based on individual health needs and risks.

4. Caloric Restriction: Caloric restriction and intermittent fasting have shown promising effects in animal studies for extending lifespan and improving healthspan. More research is needed to understand the applicability and safety of these practices in humans.

5. Exercise and Physical Activity: Regular exercise supports cardiovascular health, muscle strength, and cognitive function, contributing to healthy aging. It can also help prevent chronic diseases such as heart disease and diabetes.

6. Cognitive Stimulation: Engaging in mentally stimulating activities, such as puzzles, reading, or learning new skills, can promote cognitive health and reduce the risk of cognitive decline.

7. Disease Screening and Early Detection: Regular health check-ups and screenings for age-related diseases can lead to early detection and timely intervention, improving outcomes and disease management.

8. Vaccinations: Vaccinations, such as the flu vaccine and vaccines for specific age-related diseases like shingles, can prevent infectious diseases and their complications in older adults.

9. Medication Management: Proper management of medications is essential for older adults, as interactions and side effects can be more significant with age. Regular medication reviews and adjustments can help prevent adverse effects.

10. Personalized Medicine: Advances in personalized medicine enable tailoring treatments based on an individual's unique genetic makeup, lifestyle, and health history. This approach can improve treatment efficacy and reduce adverse effects.

11. Research and Clinical Trials: Ongoing research and clinical trials are essential for developing new anti-aging therapies and disease prevention strategies. Participation in clinical trials can contribute to scientific advancements in healthy aging.

A comprehensive approach that combines lifestyle modifications, preventive measures, and personalized interventions is the key to successful anti-aging treatments and disease prevention. Integrating these strategies can promote healthy aging, enhance well-being, and foster a society that ages with vitality and resilience.

3.2. Anti-Aging and Gene Therapies

Anti-aging and gene therapies represent cutting-edge approaches to address age-related changes and extend healthspan. Gene therapies aim to modify or repair genes associated with aging processes to promote healthier aging outcomes. While still in early stages, these innovative therapies hold promise for the future of aging research and personalized medicine.

3.2.1. The Fundamental Principles of Anti-Aging and Gene Therapies

The fundamental principles of anti-aging and gene therapies revolve around identifying and targeting key genes and biological processes associated with aging. These therapies aim to modify or repair genes to slow down the aging process, enhance cellular function, and improve overall health. By understanding the genetic basis of aging, researchers can develop targeted interventions to promote healthier aging and potentially extend lifespan.

3.2.1.1. Cellular and Molecular Gene Therapies

Cellular and molecular gene therapies are innovative approaches that hold tremendous potential in the field of anti-aging research. These therapies aim to modify or manipulate specific genes and cellular processes to promote healthier aging outcomes. Here's a detailed exploration of cellular and molecular gene therapies:

1. Gene Editing Techniques: Gene editing techniques, such as CRISPR-Cas9, allow scientists to precisely modify genes in living

cells. This revolutionary technology holds promise for correcting genetic mutations associated with aging and age-related diseases.

2. Telomere Extension: Telomeres, protective caps at the end of chromosomes, shorten with each cell division, contributing to cellular aging. Telomere extension therapies aim to elongate telomeres and promote cellular longevity.

3. Senescent Cell Clearance: Senescent cells are aged and dysfunctional cells that accumulate in tissues over time, promoting inflammation and tissue damage. Senolytic therapies target and eliminate senescent cells to improve tissue health and delay aging.

4. Stem Cell Therapies: Stem cell therapies involve using stem cells to regenerate and repair damaged tissues. These therapies hold promise for addressing age-related tissue degeneration and promoting tissue rejuvenation.

5. Epigenetic Modification: Epigenetic changes influence gene expression without altering the underlying DNA sequence. Epigenetic therapies aim to reverse age-related epigenetic modifications to restore youthful gene expression patterns.

6. Autophagy Enhancement: Autophagy is a cellular process that removes damaged components and recycles them for cellular renewal. Enhancing autophagy may improve cellular health and delay aging.

7. Mitochondrial Targeting: Mitochondria play a central role in cellular energy production and are involved in the aging process. Mitochondrial targeting therapies aim to optimize mitochondrial function and reduce oxidative stress.

8. Growth Factor Therapies: Growth factors are signaling molecules that regulate cell growth, tissue repair, and regeneration. Growth factor therapies may stimulate tissue rejuvenation and repair.

9. Nutraceutical Interventions: Nutraceuticals are bioactive compounds found in certain foods and supplements that have potential anti-aging effects. They may modulate gene expression and cellular pathways involved in aging.

10. Personalized Approaches: Cellular and molecular gene therapies are advancing toward personalized approaches. Tailoring treatments based on an individual's genetic makeup and specific aging-related challenges may optimize therapeutic outcomes.

While cellular and molecular gene therapies hold immense promise for anti-aging interventions, their development and safety require rigorous research and clinical trials. As science progresses, these therapies may revolutionize the way we address age-related changes and promote healthier aging.

3.2.1.2. Telomeres and Telomerase Therapies

Telomeres and telomerase therapies are at the forefront of anti-aging research, focusing on the manipulation of telomeres, the protective caps at the ends of chromosomes. These therapies aim to address cellular aging by maintaining or lengthening telomeres, thus promoting cellular longevity and potentially delaying the aging process. Here's a detailed exploration of telomeres and telomerase therapies:

1. Telomeres and Cellular Aging: Telomeres are repetitive DNA sequences located at the ends of chromosomes, protecting them from degradation and fusion. With each cell division, telomeres naturally shorten, limiting the cell's replicative capacity. As telomeres reach a critically short length, cells undergo senescence or apoptosis, contributing to age-related tissue decline.

2. Telomerase and Telomere Maintenance: Telomerase is an enzyme responsible for maintaining or elongating telomeres by adding telomeric DNA to the chromosome ends. It is highly active in stem cells and certain immortal cell lines but is typically suppressed in most somatic cells to prevent uncontrolled cell growth.

3. Telomerase Activation: Telomerase activation is a potential therapeutic strategy to slow down cellular aging. By reactivating telomerase in somatic cells, researchers aim to extend

the cell's replicative lifespan, potentially delaying cellular senescence and promoting tissue health.

4. Telomerase Gene Therapy: Gene therapies involving the introduction of telomerase genes into somatic cells may provide a means to activate telomerase activity and extend telomeres. This approach requires precise regulation to avoid unwanted cell proliferation, such as in cancer.

5. Telomere Lengthening Strategies: Various approaches are being explored to lengthen telomeres, including the delivery of telomere-lengthening proteins or small molecules to cells. These strategies aim to slow down the natural shortening of telomeres during cell division.

6. Potential Benefits and Risks: Telomeres and telomerase therapies offer exciting potential benefits for anti-aging interventions and regenerative medicine. By promoting cellular longevity, these therapies may contribute to improved tissue repair and healthier aging. However, concerns about the potential risks of telomerase activation, such as increased cancer risk, require careful consideration and further investigation.

7. Ethical and Safety Considerations: Developing safe and ethical telomerase therapies is paramount. Researchers must address potential ethical concerns related to the manipulation of

cellular aging and consider long-term safety implications before clinical applications.

8. Future Directions: As telomeres and telomerase continue to be the focus of anti-aging research, ongoing studies aim to elucidate the molecular mechanisms governing telomere maintenance and explore the therapeutic potential of telomerase activation in age-related conditions.

While telomeres and telomerase therapies hold immense promise for anti-aging interventions, it is essential to approach their development with caution and scientific rigor. Further research and clinical trials will determine the safety and effectiveness of these therapies, paving the way for potential breakthroughs in the pursuit of healthier aging.

3.2.2. The Potential Applications of Gene Editing and Genetic Modifications

Gene editing and genetic modifications have vast potential in medicine, agriculture, and biotechnology. They can enable personalized treatments, correct genetic diseases, develop targeted cancer therapies, enhance crops, improve biotechnology, and aid in species conservation. Responsible use and ethical considerations are essential to harness the full benefits of these technologies safely and responsibly.

3.2.2.1. Gene Editing and CRISPR-Cas9 Technology

Gene editing using CRISPR-Cas9 technology has revolutionized the field of genetics and biotechnology. CRISPR-Cas9 is a powerful and precise gene editing tool that allows researchers to modify specific DNA sequences, offering unprecedented potential in various applications. Here's a detailed exploration of gene editing and CRISPR-Cas9 technology:

1. CRISPR-Cas9 Mechanism: CRISPR-Cas9 is derived from a bacterial immune system that protects against viral infections. The CRISPR-Cas9 system uses a guide RNA to target a specific DNA sequence and the Cas9 enzyme to cut the DNA at that location.

2. Gene Editing Process: To perform gene editing with CRISPR-Cas9, scientists design a specific guide RNA that matches the target DNA sequence they want to modify. The guide RNA directs the Cas9 enzyme to the target, where it makes a precise cut in the DNA.

3. DNA Repair Mechanisms: After the DNA is cut, the cell's natural repair mechanisms come into play. There are two primary repair pathways: Non-Homologous End Joining (NHEJ) and Homology-Directed Repair (HDR).

4. Non-Homologous End Joining (NHEJ): NHEJ is an error-prone repair process that often results in small insertions or

deletions in the DNA sequence. These alterations can lead to gene knockout or inactivation.

5. Homology-Directed Repair (HDR): HDR is a more precise repair mechanism that uses a template DNA sequence to fix the cut. This pathway allows for targeted insertion or replacement of specific DNA sequences.

6. Applications in Medicine: CRISPR-Cas9 has tremendous potential in treating genetic diseases. By editing disease-causing mutations, researchers aim to correct genetic disorders at their source.

7. Cancer Therapies: CRISPR-Cas9 can be used to target cancer-related genes, potentially disrupting tumor growth and enhancing the body's immune response against cancer cells.

8. Agriculture and Food Security: CRISPR-Cas9 can improve crops by making them more resistant to pests and diseases, increasing their nutritional value, and enhancing overall crop yield for global food security.

9. Biotechnology and Drug Development: CRISPR-Cas9 plays a critical role in drug discovery and development, enabling researchers to study the function of genes and identify potential therapeutic targets.

10. Ethical Considerations: While CRISPR-Cas9 technology offers transformative possibilities, ethical concerns surround its use in human germline editing and the potential for unintended off-target effects.

CRISPR-Cas9 technology continues to advance rapidly, offering immense potential in diverse fields. As research and technology progress, responsible use and ethical oversight remain paramount to harness the full potential of gene editing for the benefit of humanity.

3.2.2.2. Genetic Modifications and Anti-Aging Treatments

Genetic modifications hold promising implications for anti-aging treatments. By targeting specific genes and cellular pathways associated with aging, scientists aim to develop interventions that slow down the aging process and promote healthier aging outcomes. Here's a detailed exploration of genetic modifications and their potential in anti-aging treatments:

1. Telomere Lengthening and Senescent Cell Clearance: Genetic modifications can target telomeres and senescent cells to mitigate cellular aging. Lengthening telomeres and clearing senescent cells may delay tissue degeneration and improve overall healthspan.

2. Cellular Rejuvenation and Regeneration: Genetic modifications might enable the activation of stem cells and promote tissue regeneration, rejuvenating aged or damaged tissues and organs.

3. Epigenetic Reversal and Gene Expression Regulation: Epigenetic modifications can influence gene expression patterns associated with aging. Genetic modifications might reverse age-related epigenetic changes to restore youthful gene expression profiles.

4. Antioxidant and Detoxification Pathways: Enhancing antioxidant defenses and detoxification pathways through genetic modifications may reduce oxidative stress and the accumulation of cellular damage, potentially slowing down aging.

5. Immune System Enhancement: Genetic modifications could strengthen the immune system to better combat infections, prevent chronic inflammation, and improve immune surveillance against cancer.

6. Mitochondrial Optimization: Mitochondria play a vital role in cellular energy production and aging. Genetic modifications may optimize mitochondrial function and reduce mitochondrial dysfunction associated with aging.

7. Disease Resistance and Resilience: Genetic modifications might confer resistance to age-related diseases, making individuals less susceptible to certain health conditions as they age.

8. Personalized Anti-Aging Interventions: Genetic modifications enable personalized anti-aging interventions based on an individual's genetic makeup and specific aging-related challenges.

9. Ethical and Safety Considerations: As with any genetic intervention, ethical considerations and safety protocols are essential to avoid unintended consequences and ensure responsible use.

10. Future Challenges and Opportunities: While genetic modifications offer exciting possibilities for anti-aging treatments, extensive research, clinical trials, and regulatory oversight are necessary to realize their potential and address potential risks.

As science advances, genetic modifications may play a crucial role in the development of tailored and effective anti-aging therapies. However, the responsible use of these technologies and ethical considerations must accompany these advancements to ensure their safe and effective application in promoting healthier aging and extending healthspan.

3.2.3. The Social and Ethical Dimensions of Anti-Aging Treatments

Anti-aging treatments raise significant social and ethical considerations. As these technologies advance, it becomes essential to address concerns related to accessibility, equity, informed consent, and potential unintended consequences. The ethical implications of altering the aging process and extending human lifespan require careful examination to ensure responsible and equitable use of these transformative interventions.

3.2.3.1. Aging and Social Inequalities

Aging and social inequalities are deeply interconnected, presenting various challenges and disparities for older adults across different socioeconomic backgrounds. Here's a detailed exploration of the relationship between aging and social inequalities:

1. Healthcare Access and Quality: Socioeconomic status often affects access to healthcare services and the quality of care received. Lower-income individuals may face barriers to essential medical resources, preventive screenings, and specialized treatments, leading to disparities in health outcomes as they age.

2. Economic Security and Retirement: Economic disparities impact retirement preparedness and financial security in old age. Those with limited financial resources may experience greater

difficulties in maintaining a comfortable standard of living during retirement, leading to financial stress and hardships.

3. Housing and Living Conditions: Inadequate housing and living conditions disproportionately affect older adults with lower incomes. Limited access to suitable housing, safe neighborhoods, and supportive communities can negatively impact their health and well-being.

4. Social Isolation and Loneliness: Social isolation and loneliness can be more prevalent among older adults with limited social networks and fewer resources. Lack of access to community activities and social support can have adverse effects on mental health and overall life satisfaction.

5. Long-Term Care Services: Quality long-term care services are crucial for older adults with complex health needs. Social inequalities may hinder access to affordable and high-quality long-term care, impacting the ability of older adults to age in a dignified and supported manner.

6. Digital Divide: With technological advancements, the digital divide becomes a significant issue for older adults. Those with limited access to technology may face challenges in accessing online services, information, and healthcare resources.

7. Healthcare Decision-Making and Advocacy: Disparities in education and healthcare literacy may influence older adults' ability to make informed decisions about their health and advocate for their needs within the healthcare system.

8. Social Services and Support Programs: Access to social services and support programs, such as home care assistance or meal delivery, may vary based on socioeconomic status, impacting the overall well-being and independence of older adults.

9. Ageism and Stereotypes: Ageism and negative stereotypes about aging can exacerbate social inequalities, leading to biased treatment and reduced opportunities for older adults.

Addressing social inequalities in aging requires a comprehensive approach involving policymakers, healthcare providers, community organizations, and society as a whole. Strategies to promote equitable access to healthcare, social services, and economic opportunities can contribute to a more inclusive and supportive environment for older adults, ensuring that aging is a positive and fulfilling experience regardless of one's socioeconomic background.

3.2.3.2. Accessibility and Justice Issues of Anti-Aging Treatments

As anti-aging treatments advance, issues of accessibility and justice become critical considerations. Ensuring equitable access to these therapies and addressing potential ethical concerns are essential to avoid widening existing health disparities. Here's a detailed exploration of the accessibility and justice issues related to anti-aging treatments:

1. Affordability and Cost: Anti-aging treatments may initially be expensive and limited to those with substantial financial means. This could create a two-tiered healthcare system, where only the wealthy can access cutting-edge therapies, exacerbating social inequalities.

2. Disparities in Research and Development: Disparities in research funding and priorities may lead to the development of anti-aging treatments that primarily address the health concerns of certain populations, neglecting the needs of marginalized or underserved groups.

3. Access to Information and Informed Consent: Access to accurate information about anti-aging treatments can vary, leading to potential misinformation and inadequate informed consent. Ensuring that individuals have access to comprehensive and unbiased information is vital for making informed decisions about their health.

4. Vulnerable Populations: Vulnerable populations, such as the elderly, people with disabilities, and minority groups, may face

additional barriers in accessing anti-aging treatments due to systemic biases and discrimination.

5. Equitable Distribution of Benefits: The benefits of anti-aging treatments should be distributed fairly across different demographics. Ensuring that these therapies benefit a wide range of individuals is crucial for promoting social justice and equity.

6. Long-Term Safety and Risks: Monitoring the long-term safety and potential risks of anti-aging treatments is essential, as unforeseen side effects could disproportionately impact vulnerable populations.

7. Ethical Considerations: Ethical dilemmas, such as germline editing, require careful consideration to prevent unintended consequences and uphold the principles of justice and fairness.

8. Global Access: Ensuring global access to anti-aging treatments is essential to avoid creating a divide between countries with differing resources and healthcare systems.

9. Regulatory Oversight: Effective regulatory frameworks are necessary to ensure that anti-aging treatments undergo rigorous evaluation for safety and efficacy, preventing the proliferation of unproven or potentially harmful interventions.

10. Collaboration and Advocacy: Collaboration between policymakers, healthcare professionals, patient advocates, and the public is essential for addressing accessibility and justice issues in anti-aging treatments. Advocacy for equitable access and justice-driven policies can promote positive change.

Addressing accessibility and justice issues is integral to harnessing the benefits of anti-aging treatments for the betterment of society. By ensuring equitable access and promoting social justice principles, we can work towards a future where anti-aging interventions are available to all, regardless of their socioeconomic background or geographic location.

3.3. The Societal and Economic Impacts of Longevity

Longer life spans have far-reaching effects on societies and economies. Healthcare systems, retirement patterns, workforce dynamics, and family structures undergo changes to accommodate an aging population. Adjustments in policies and innovations become essential to support the well-being and productivity of older adults and sustain economic growth.

3.3.1. The Demographic and Economic Consequences of Longevity

Longevity's demographic and economic consequences are profound. A growing elderly population influences the dependency ratio, putting pressure on healthcare and social welfare systems. Additionally, an extended workforce participation may boost economic productivity, but it necessitates pension and retirement adjustments. Balancing these factors is crucial for maintaining societal well-being and economic sustainability.

3.3.1.1. The Impact of Extended Longevity on Population Structure

Extended longevity significantly impacts the population structure, leading to demographic shifts and changing age distributions. Here's a detailed exploration of the impact of extended longevity on population structure:

Aging Population: With increased life expectancy, a larger proportion of the population comprises older adults. This demographic shift results in an aging population, characterized by a higher median age.

Dependency Ratio: The aging population affects the dependency ratio, which measures the number of non-working individuals (young and elderly) relative to the working-age population. A higher dependency ratio places financial strain on the workforce to support social welfare and healthcare for the elderly.

Challenges for Social Welfare Systems: The rising number of older adults places greater demands on social welfare systems, including pension plans, healthcare, and long-term care services. Governments and policymakers face challenges in ensuring the sustainability and adequacy of these support systems.

Healthcare Demand: An aging population increases the demand for healthcare services, especially for age-related conditions and chronic diseases. Healthcare systems must adapt to cater to the unique needs of elderly patients and promote healthy aging.

Labor Market Impact: An older workforce may remain employed for longer, contributing to economic productivity. However, it may also create potential challenges for younger individuals seeking job opportunities and career advancement.

Silver Economy: The growing number of older consumers drives the emergence of the "silver economy," with businesses catering to the unique preferences and needs of the elderly market.

Pension and Retirement Reforms: Extended longevity necessitates adjustments in pension and retirement policies to ensure the financial security of retirees throughout their extended lifespans.

Inter-Generational Relationships: A shift in population structure affects inter-generational relationships and family dynamics. Younger generations may take on caregiving roles and experience changes in financial responsibilities.

Policy Implications: Policymakers must address the implications of an aging population on various sectors, including healthcare, labor, social welfare, and economic planning. Strategies for promoting active aging and inter-generational cohesion are essential.

Global Perspectives: Extended longevity impacts countries worldwide, with some regions experiencing more pronounced demographic shifts. International collaboration is crucial for sharing knowledge and best practices in addressing the challenges of an aging population.

Addressing the impact of extended longevity on population structure requires a multi-faceted approach, encompassing policy reforms, healthcare advancements, and fostering age-friendly communities. Proactive measures can help societies harness the benefits of longevity while mitigating potential challenges for the well-being and prosperity of all age groups.

3.3.1.2. Longevity's Effects on Retirement and Working Life

Extended longevity has significant effects on retirement patterns and working life, transforming how individuals approach their careers and plan for their later years. Here's a detailed exploration of longevity's impact on retirement and working life:

1. Delayed Retirement Age: Longer life expectancies lead to extended working lives, as individuals seek to maintain financial security throughout their longer retirement periods. Many choose to delay retirement and remain in the workforce for more years.

2. Flexible Retirement Options: With longer lifespans, individuals may seek flexible retirement arrangements, such as gradual retirement or part-time work, allowing for a smoother transition from full-time employment to full retirement.

3. Retirement Planning Challenges: Extended retirement periods pose challenges in financial planning. Individuals need to accumulate more significant savings to sustain their lifestyles during extended retirement years.

4. Pension System Adjustments: Pension systems may require adjustments to accommodate longer retirement periods. Governments and employers may consider revised pension plans to ensure adequacy and fairness.

5. Skills and Knowledge Transfer: Older workers' extended participation in the workforce allows for valuable skills and knowledge transfer to younger generations, contributing to enhanced productivity and a skilled labor force.

6. Age Diversity in the Workplace: The aging workforce brings age diversity to workplaces, promoting inter-generational collaboration and creating a more inclusive work environment.

7. Workforce Management Strategies: Employers may implement workforce management strategies to support older employees, including age-sensitive policies, training programs, and ergonomic accommodations.

8. Redefining Career Trajectories: Longer working lives offer opportunities for career shifts and redefining career trajectories, enabling individuals to pursue new interests or transition into less physically demanding roles.

9. Retirement Savings and Investments: Extended lifespans necessitate more significant emphasis on retirement savings and investment planning to maintain financial stability throughout retirement.

10. Work-Life Balance: As individuals work longer, maintaining work-life balance becomes crucial for overall well-being. Flexible work arrangements and supportive workplace policies can promote work-life harmony.

Longevity's impact on retirement and working life presents a complex set of challenges and opportunities. Governments, employers, and individuals must collaborate to develop age-inclusive policies, foster career adaptability, and ensure that longer working lives contribute to personal fulfillment and societal progress.

3.3.2. The Societal and Cultural Impacts of Longevity

Longevity's societal and cultural impacts are far-reaching. Longer life spans influence family dynamics, inter-generational relationships, caregiving roles, and the perception of aging. These changes prompt cultural shifts in how societies view aging, retirement, and the role of older adults in community life. Embracing and adapting to these impacts are essential for fostering age-friendly societies and promoting positive attitudes towards aging.

3.3.2.1. Aging and Changes in Family Structure

Longevity significantly influences family structures, as extended lifespans reshape inter-generational relationships and caregiving dynamics. Here's a detailed exploration of the impact of aging on family structures:

Multi-Generational Households: Longer life expectancies lead to multi-generational households, with older adults living alongside their adult children and grandchildren. This fosters close family ties and provides mutual support.

Role Reversal: Aging parents may require care and assistance, leading to role reversals where adult children take on caregiving responsibilities for their elderly parents.

Caregiving Challenges: The increasing number of older adults places greater demands on family caregivers, who may need to balance work, family, and caregiving responsibilities.

Geographical Proximity: Extended lifespans may prompt older adults to move closer to their adult children to receive care and support, strengthening geographical proximity among family members.

Financial Support: Longer life expectancies may result in older adults living longer in retirement, potentially requiring financial support from their adult children.

Marriage and Repartnering in Later Life: With longer life expectancies, some older adults may enter new relationships or remarry later in life, influencing family structures and dynamics.

Grandparenting Roles: Longer life spans provide opportunities for grandparents to be actively involved in the lives of their grandchildren, contributing to the social and emotional development of younger generations.

Estate Planning and Inheritance: Extended lifespans may necessitate careful estate planning and inheritance arrangements to ensure financial stability and equitable distribution of assets.

Intergenerational Cohesion: As families navigate the challenges and opportunities of aging, fostering intergenerational cohesion becomes essential for building supportive and connected family units.

Cultural Norms and Traditions: Cultural norms around aging and family dynamics may evolve as societies adapt to the changing demographic landscape.

Understanding the impact of aging on family structures is crucial for developing supportive policies and services that address the needs of older adults and their families. Strengthening inter-generational relationships and providing adequate resources for family caregivers can promote family well-being and enhance the quality of life for all family members across generations.

3.3.2.2. Longevity's Effects on Society and Culture Dynamics

Longevity's effects on society and culture are multifaceted, influencing attitudes, values, and traditions surrounding aging and the role of older adults. Here's a detailed exploration of how longevity impacts society and culture dynamics:

Changing Perceptions of Aging: Longer life expectancies challenge traditional notions of aging as a time of decline. Societal perceptions may shift to view older adults as valuable contributors with diverse experiences and wisdom.

Productive Aging: Extended lifespans enable older adults to remain active and engaged in various roles, such as the workforce, volunteering, and community leadership, contributing to society's productivity.

Cultural Norms and Roles: As life spans increase, cultural norms regarding retirement, family obligations, and inter-generational relationships may undergo transformation.

Intergenerational Cohesion: Longer life expectancies promote inter-generational relationships, fostering stronger connections and mutual support between older adults and younger generations.

Impact on Workforce: An aging population may necessitate policy changes to promote age-inclusive workplaces, accommodate older workers' needs, and harness their skills and experience.

Arts and Media Representation: Longer life spans can influence how older adults are portrayed in arts and media, potentially promoting positive and diverse depictions of aging.

Health and Wellness Culture: The focus on healthy aging may lead to the emergence of a health and wellness culture that emphasizes preventive healthcare and lifestyle choices conducive to long-term well-being.

Age-Friendly Communities: Societies may prioritize age-friendly community planning to create environments that cater to the needs and preferences of older residents.

Longevity Celebrations: Cultures may develop unique celebrations or rituals to honor and celebrate longevity milestones, promoting a positive outlook on aging.

Cultural Exchange and Global Perspectives: Longer life spans foster cultural exchange and knowledge-sharing about successful aging practices across diverse societies and regions.

Elderly Advocacy and Empowerment: Increased focus on older adults' well-being may lead to greater advocacy for their rights, well-being, and empowerment within society.

Cultural Adaptation: Societies may adapt cultural practices and traditions to accommodate the needs and preferences of older adults, promoting inclusivity and respect for all age groups.

Addressing longevity's effects on society and culture dynamics involves fostering age-inclusive policies, challenging ageist stereotypes, and recognizing the valuable contributions of older adults to societal well-being. Embracing and celebrating the diversity of aging experiences can create more vibrant and harmonious societies that benefit people of all ages.

3.3.3. The Impact of Longevity on Healthcare Systems and Lifestyle

Longevity's impact on healthcare systems and lifestyle is profound. Longer life spans necessitate adjustments in healthcare delivery to meet the evolving needs of an aging population. Individuals are encouraged to adopt healthier lifestyles to promote active aging and maintain their well-being throughout their extended lifespans. Emphasizing preventive care and promoting age-friendly healthcare services are essential for supporting the health and quality of life of older adults.

3.3.3.1. The Effect of Longevity on Healthcare Expenses

Longevity has a significant effect on healthcare expenses, as longer life spans lead to increased demand for healthcare services, age-related treatments, and long-term care. Here's a detailed exploration of the impact of longevity on healthcare expenses:

Age-Related Diseases: Longer life expectancies result in a higher prevalence of age-related diseases such as cardiovascular conditions, neurodegenerative disorders, and certain cancers, leading to increased healthcare utilization and expenses.

Chronic Conditions: Older adults often have multiple chronic health conditions, requiring ongoing medical management and specialized care, which contributes to higher healthcare costs.

Long-Term Care Needs: An aging population requires more long-term care services, including nursing homes, assisted living facilities, and home care, adding to healthcare expenses.

Medical Technology and Treatments: Advances in medical technology and treatments for age-related conditions may improve health outcomes but could also escalate healthcare expenses.

End-of-Life Care: Longer life spans mean more individuals will require end-of-life care, which can be resource-intensive and costly.

Healthcare Workforce Demands: The aging population places increased demands on healthcare professionals, leading to potential workforce shortages and increasing labor costs.

Health Insurance and Coverage: Healthcare systems may face challenges in providing adequate health insurance and coverage for a growing elderly population, potentially affecting healthcare accessibility.

Preventive Care: Emphasizing preventive care can help reduce healthcare expenses by addressing health issues early and preventing the progression of chronic conditions.

Health Promotion and Education: Encouraging healthy lifestyles and promoting public health initiatives can lead to better health outcomes and potentially reduce the burden on healthcare systems.

Policy Interventions: Governments and policymakers may need to implement healthcare reforms and policies to ensure the sustainability and affordability of healthcare for an aging population.

Addressing the effect of longevity on healthcare expenses requires a comprehensive approach, involving preventive measures, targeted interventions for chronic conditions, and investments in healthcare infrastructure and workforce. Striking a balance between ensuring quality healthcare for older adults while managing costs is crucial for maintaining a sustainable and inclusive healthcare system.

3.3.3.2. Longevity's Influence on Healthcare and Patient Care

Longevity profoundly influences healthcare and patient care, as healthcare systems adapt to the needs of an aging population. Here's a detailed exploration of how longevity impacts healthcare and patient care:

Geriatric Medicine Specialization: Longer life spans necessitate a focus on geriatric medicine, with healthcare professionals specializing in the unique healthcare needs of older adults.

Age-Related Conditions: Healthcare providers must address age-related conditions, such as osteoporosis, dementia, and mobility issues, requiring specialized treatment and care plans.

Chronic Disease Management: As older adults often have multiple chronic conditions, effective disease management becomes crucial to maintain their health and quality of life.

Integrated Care: Healthcare systems may adopt integrated care models to ensure seamless coordination between various healthcare providers and services involved in managing complex health conditions.

Holistic Approach: Patient care for older adults often requires a holistic approach, considering not only physical health but also mental, emotional, and social well-being.

Preventive Care: Emphasizing preventive care becomes essential to promote healthy aging, detect potential health issues early, and reduce the need for costly treatments.

Telemedicine and Remote Monitoring: Telemedicine and remote monitoring technologies allow healthcare professionals to reach older adults in remote areas and provide continuous monitoring of chronic conditions.

Patient-Centered Care: Healthcare systems prioritize patient-centered care, ensuring that medical decisions align with the preferences and values of older patients.

Palliative and End-of-Life Care: As life spans increase, palliative care becomes increasingly important to enhance the quality of life for older adults facing serious illnesses.

Caregiver Support: Healthcare systems may offer caregiver support programs to assist family members in caring for older adults and managing their health needs.

Health Literacy and Education: Promoting health literacy and patient education helps older adults make informed decisions about their health and actively participate in their care.

Age-Friendly Facilities: Healthcare facilities may adopt age-friendly design principles to create a supportive and comfortable environment for older patients.

Longevity's influence on healthcare and patient care requires healthcare systems to adapt, innovate, and prioritize the unique needs of older adults. By focusing on preventive care, patient-centered approaches, and specialized geriatric services, healthcare providers can optimize the health outcomes and overall well-being of the aging population.

CHAPTER 4

Superhumans and Bioengineering

4.1. Genetic Modification and Human Enhancement

Genetic modification refers to the alteration of an organism's genetic material, and it has the potential to be a powerful tool for human enhancement. Through gene editing technologies like CRISPR-Cas9, scientists can target and modify specific genes, raising ethical and moral questions about the boundaries of human enhancement and the potential consequences of manipulating our genetic makeup to improve physical and cognitive traits. Discussions around genetic modification and human enhancement delve into issues of equity, consent, and the potential for unintended consequences, necessitating careful consideration as we navigate this rapidly advancing field.

4.1.1. Genetic Engineering and the Editing of the Human Genome

Genetic engineering involves deliberate alterations to an organism's DNA, including the editing of the human genome. Technologies like CRISPR-Cas9 enable precise modifications, raising ethical debates about the potential benefits and risks of gene editing in humans. While it holds promise for addressing genetic diseases, concerns over unintended consequences and ethical boundaries underscore the need for responsible and cautious use of genetic engineering in the context of the human genome.

4.1.1.1. The History and Evolution of Genetic Modification

The history of genetic modification traces back to ancient agricultural practices, but it has evolved significantly with advances in science and technology. Here's a detailed exploration of the history and evolution of genetic modification:

1. Early Beginnings: The earliest form of genetic modification can be traced back to the domestication of plants and animals by ancient agricultural societies around 10,000 years ago. Selective breeding allowed humans to cultivate desired traits in crops and livestock.

2. Discovery of DNA: The understanding of genetics took a significant leap forward with the discovery of the structure of DNA by James Watson and Francis Crick in 1953. This breakthrough laid the foundation for modern genetic research.

3. Recombinant DNA Technology: In the 1970s, the development of recombinant DNA technology enabled scientists to

combine DNA from different organisms, paving the way for genetic engineering.

4. First Genetically Modified Organism (GMO): In 1973, Herbert Boyer and Stanley Cohen successfully created the first genetically modified organism, a bacterium that incorporated foreign DNA.

5. Transgenic Animals and Plants: In the 1980s, researchers began to genetically engineer animals and plants to introduce desirable traits, such as pest resistance and improved nutritional content.

6. Human Genome Project: The completion of the Human Genome Project in 2003 marked a milestone in understanding the human genome's structure and function, providing essential knowledge for genetic research.

7. CRISPR-Cas9 Revolution: The discovery of CRISPR-Cas9 in 2012 revolutionized genetic engineering, offering a precise and efficient method for editing DNA. Its applications in research and potential therapeutic uses garnered significant attention.

8. Gene Therapy Advancements: Gene therapy, a medical application of genetic modification, has made significant strides in treating genetic disorders and other diseases by targeting and correcting faulty genes.

9. Ethical and Regulatory Debates: As genetic modification advanced, ethical and regulatory debates intensified, particularly regarding the use of gene editing in humans, with concerns about unintended consequences and the potential for germline modifications.

10. Potential and Future Prospects: Genetic modification holds promise for improving agriculture, healthcare, and environmental conservation. Research continues on therapies for genetic diseases and potential applications in human enhancement.

The history and evolution of genetic modification reflect human ingenuity and scientific progress. While it presents immense possibilities for addressing challenges, it also requires careful consideration of ethical, social, and environmental implications to ensure responsible and beneficial use.

4.1.1.2. Genetic Editing and Gene Alteration Techniques

Genetic editing encompasses a variety of techniques that allow scientists to modify an organism's DNA, including the human genome. These techniques hold great potential for medical advancements, but they also raise ethical questions and concerns about unintended consequences. Here's a detailed exploration of genetic editing and gene alteration techniques:

1. Recombinant DNA Technology: Recombinant DNA technology involves combining DNA from different sources to create new genetic sequences. This technique laid the groundwork for genetic engineering.

2. Gene Knockout: Gene knockout is a technique where specific genes are intentionally deactivated or "knocked out" in an organism's genome to study their function or assess their role in diseases.

3. Gene Addition: Gene addition involves inserting new genetic material into an organism's DNA, typically to compensate for faulty genes responsible for genetic disorders.

4. Gene Replacement: Gene replacement aims to replace a faulty gene with a healthy, functional version to correct genetic mutations associated with diseases.

5. Site-Directed Mutagenesis: Site-directed mutagenesis is a technique that allows scientists to make precise changes in a gene's sequence, introducing or correcting specific mutations.

6. CRISPR-Cas9: CRISPR-Cas9 is a revolutionary gene editing tool derived from a bacterial defense system. It uses RNA molecules to guide the Cas9 enzyme to a specific DNA sequence, where it can create double-stranded breaks, leading to targeted gene modifications.

7. TALEN (Transcription Activator-Like Effector Nuclease): TALENs are enzymes that function similarly to CRISPR-Cas9 but use a different DNA-targeting mechanism to induce gene modifications.

8. ZFN (Zinc Finger Nuclease): ZFNs are engineered proteins that can recognize and bind to specific DNA sequences, leading to targeted gene editing.

9. HDR (Homology-Directed Repair): HDR is a DNA repair mechanism that can be harnessed in gene editing to introduce specific changes at targeted genomic sites.

10. Non-Homologous End Joining (NHEJ): NHEJ is another DNA repair mechanism used in gene editing, often resulting in small insertions or deletions that may disrupt gene function.

The advancement of genetic editing and gene alteration techniques has opened up new possibilities for treating genetic diseases, enhancing agricultural productivity, and exploring fundamental biological processes. However, the ethical implications of manipulating the human genome, particularly germline modifications, have sparked ethical debates worldwide. The responsible and ethical use of these powerful techniques remains a critical consideration as we navigate the frontiers of genetic editing and its potential impact on humanity.

4.1.2. The Potential Impact of Genetic Modification on Health and Performance

Genetic modification holds the potential to significantly impact human health and performance. By targeting specific genes associated with disease susceptibility and physical attributes, it may offer promising avenues for disease prevention, treatment, and human enhancement. However, careful ethical considerations and thorough scientific research are essential to ensure the responsible and safe use of genetic modification technologies in these domains.

4.1.2.1. The Effects of Genetic Modifications on Genetic Diseases and Inherited Predispositions

Genetic modifications have the potential to revolutionize the treatment and prevention of genetic diseases and inherited predispositions. Here's a detailed exploration of the effects of genetic modifications in these areas:

1. Targeted Gene Therapy: Genetic modifications can be used in gene therapy to target and correct specific genetic mutations responsible for inherited diseases. This approach holds promise for treating a wide range of genetic disorders.

2. Disease Prevention: Genetic modifications can help prevent the transmission of certain genetic diseases from parents

to their offspring by correcting or eliminating disease-causing mutations in the germline.

3. Precision Medicine: Genetic modifications allow for personalized treatment strategies tailored to an individual's genetic makeup, enhancing the effectiveness of medical interventions.

4. Ethical Considerations: The use of genetic modifications for disease prevention raises ethical questions about the extent to which we should intervene in the human germline and the potential long-term consequences of such interventions.

5. Enhanced Disease Understanding: By studying the effects of genetic modifications in laboratory settings, researchers gain valuable insights into the genetic basis of various diseases, leading to better diagnostic and treatment approaches.

6. Reduced Disease Burden: Successful genetic modifications could significantly reduce the burden of genetic diseases on individuals, families, and healthcare systems.

7. Gene Editing Challenges: While gene editing offers potential therapeutic benefits, technical challenges, off-target effects, and unintended consequences require careful consideration.

8. Inherited Predispositions: Genetic modifications may help mitigate the impact of inherited predispositions to certain diseases, offering a potential avenue for preventive medicine.

9. Gene Regulation and Expression: Genetic modifications can influence gene regulation and expression, providing opportunities to fine-tune biological processes and address disease-related dysfunctions.

10. Future Directions: Ongoing research continues to explore the possibilities and limitations of genetic modifications in treating genetic diseases and managing inherited predispositions.

Balancing the potential benefits of genetic modifications in addressing genetic diseases and inherited predispositions with the ethical considerations and scientific challenges is critical in harnessing these technologies for the betterment of human health. Transparency, informed consent, and rigorous oversight are essential components of responsible genetic research and application in medical contexts.

4.1.2.2. Genetic Modification and Sports and Performance Abilities

Genetic modification has garnered attention in the realm of sports and performance enhancement, with the potential to influence athletic abilities. Here's a detailed exploration of the relationship between genetic modification and sports and performance abilities:

1. Gene Variants and Athletic Performance: Certain gene variants are associated with athletic performance traits, such as muscle strength, endurance, oxygen utilization, and recovery. Genetic modifications could potentially target these genes to enhance specific athletic traits.

2. Myostatin Inhibition: Myostatin is a protein that regulates muscle growth. Inhibiting myostatin through genetic modifications could theoretically increase muscle mass and strength, leading to potential performance improvements.

3. Erythropoietin (EPO) Production: EPO is a hormone that stimulates red blood cell production and enhances oxygen-carrying capacity. Genetic modifications could potentially increase EPO production, improving endurance and stamina.

4. Precision Training: Genetic modifications may aid in precision training by identifying genetic variants associated with optimal training responses, injury risk, and recovery.

5. Ethical Considerations: The use of genetic modifications in sports raises ethical concerns about fairness, safety, and the potential for creating unequal advantages among athletes.

6. Regulatory Restrictions: Current regulations in sports prohibit the use of genetic modifications for performance enhancement. However, the evolving landscape of genetic research necessitates ongoing discussions about the boundaries of fair play and athlete safety.

7. Potential Misuse: Concerns exist about the misuse of genetic modifications in sports, leading to a potential arms race of genetic enhancements and compromising the integrity of competitions.

8. Long-Term Health Impact: The long-term health effects of genetic modifications for performance enhancement are not yet fully understood, necessitating careful evaluation and risk assessment.

9. Collaborative Efforts: The sports community, researchers, and governing bodies must collaborate to establish guidelines and regulations that strike a balance between ethical considerations and the advancement of scientific knowledge.

10. Potential Benefits for Rehabilitation: Genetic modifications may also hold promise in rehabilitation scenarios, aiding athletes in recovering from injuries more effectively.

Genetic modification's potential influence on sports and performance abilities raises complex ethical, scientific, and social implications. While the technology holds promise for addressing genetic diseases and enhancing human health, its application in the context of sports requires careful scrutiny and thoughtful deliberation to ensure fair play, athlete safety, and the preservation of the spirit of sport.

4.1.3. The Ethical and Legal Dimensions of Genetic Modification

Genetic modification raises profound ethical and legal considerations. Ethical dilemmas center on the boundaries of human enhancement, consent, equity, and potential unintended consequences. Legal dimensions involve regulations around gene editing, including the use of genetic modifications in medicine and sports, ensuring responsible and accountable practices. Striking the right balance between scientific advancements, individual rights, and societal well-being is crucial as we navigate the complex landscape of genetic modification.

4.1.3.1. The Distinction Between Natural and Artificial Changes in Genetic Modification

The distinction between natural and artificial changes in genetic modification lies at the heart of ethical debates surrounding this technology. Here's a detailed exploration of this distinction:

1. Natural Changes: In nature, genetic changes occur through natural processes like mutations and genetic recombination. These changes shape the genetic diversity of living organisms and contribute to evolution.

2. Artificial Changes: Genetic modification introduces deliberate alterations to an organism's DNA using human-made technologies like CRISPR-Cas9. These changes are directed and purposeful.

3. Ethical Considerations: The distinction between natural and artificial changes raises ethical questions about human intervention in the genetic makeup of living beings. While natural changes are seen as part of the natural course of evolution, artificial changes may be viewed as human interference in the natural order.

4. Intention and Control: Artificial changes in genetic modification are guided by human intentions and goals, leading to discussions about the ethical implications of manipulating the genetic characteristics of living organisms.

5. Unintended Consequences: Artificial changes may carry the risk of unintended consequences, such as off-target effects or unforeseen genetic mutations. Ethical considerations involve the need to balance potential benefits with potential risks.

6. Informed Consent: In human gene editing and therapies, informed consent becomes essential to ensure individuals understand the potential consequences of genetic modifications and exercise autonomy over their genetic information.

7. Scope of Application: The distinction between natural and artificial changes affects the scope of genetic modification applications, guiding ethical and legal regulations around its use in medicine, agriculture, and other fields.

8. Germline Editing: Germline editing, which involves modifying the DNA of reproductive cells, blurs the line between natural and artificial changes, leading to significant ethical debates about the potential long-term effects on future generations.

9. Cultural and Societal Perspectives: Perspectives on the distinction between natural and artificial changes may vary across cultures and societies, influencing attitudes toward genetic modification and its acceptance.

Navigating the distinction between natural and artificial changes in genetic modification requires comprehensive ethical assessments, public dialogue, and adherence to legal frameworks. Striking a balance between the potential benefits of genetic modification and respect for the integrity of natural systems is crucial in shaping responsible and ethical practices in this rapidly advancing field.

4.1.3.2. The Ethics of Genetic Modification and Human Rights Considerations

The ethics of genetic modification intersect with human rights considerations, presenting complex ethical dilemmas that necessitate thoughtful examination. Here's a detailed exploration of the ethics of genetic modification and its implications for human rights:

1. Autonomy and Informed Consent: The principle of autonomy emphasizes individuals' right to make informed decisions about their bodies and genetic information. In the context of genetic modification, obtaining informed consent becomes crucial to respect individuals' autonomy and agency.

2. Non-Discrimination and Equity: Genetic modifications that lead to enhancements raise concerns about creating unequal advantages among individuals. Ensuring equitable access to genetic therapies and preventing discrimination based on genetic traits are essential human rights considerations.

3. Privacy and Genetic Data Protection: As genetic information becomes more accessible, safeguarding privacy and protecting genetic data against misuse and unauthorized access become imperative to preserve individuals' rights to privacy and autonomy.

4. Germline Editing and Future Generations: Germline editing, which alters the genetic makeup of reproductive cells, raises ethical questions about the long-term implications for future generations. Balancing the potential benefits with the potential risks and unintended consequences becomes crucial in ethical decision-making.

5. Societal and Cultural Perspectives: Societal and cultural perspectives on genetic modification may vary, influencing ethical discussions and policy development. Respect for diverse viewpoints is essential in addressing human rights considerations related to genetic modification.

6. Justice and Access: Ensuring justice involves equitable access to genetic therapies and interventions. Genetic modifications must be available and affordable to all, regardless of socio-economic status, to uphold human rights principles of justice and fairness.

7. Unintended Consequences and Risks: Ethical considerations involve evaluating the potential risks and unintended consequences of genetic modifications, aiming to minimize harm to individuals and society.

8. Procreative Liberty: Genetic modification raises questions about procreative liberty, allowing individuals to make

decisions about their reproductive choices while acknowledging the broader societal implications.

9. Public Engagement and Transparency: The ethical dimensions of genetic modification call for public engagement, transparency, and open dialogue to ensure collective decision-making and accountability.

10. International Human Rights Framework: Applying a human rights lens to genetic modification involves adhering to international human rights frameworks and ensuring that advancements in genetic technologies align with fundamental human rights principles.

Balancing the ethical considerations of genetic modification with human rights requires interdisciplinary collaboration among researchers, policymakers, ethicists, and the broader public. Upholding human rights principles while harnessing the potential of genetic modification for beneficial applications is a fundamental responsibility in shaping the ethical trajectory of this field.

4.2. Superhumans and Advanced Bioengineering

The concept of "superhumans" and advanced bioengineering envisions the enhancement of human capabilities beyond natural limits through cutting-edge technologies. This fascinating area explores the potential to elevate physical and cognitive traits, sparking discussions about ethical implications, societal impact, and the definition of humanity. While the idea captivates scientific imagination, it also raises profound questions about the responsible and ethical use of bioengineering in shaping the future of humanity.

4.2.1. The Concept of Superhumans and Enhanced Abilities

The concept of superhumans revolves around the idea of using advanced bioengineering to elevate human capabilities beyond natural limits. It involves enhancing physical strength, endurance, cognitive functions, and other traits, sparking debates about the ethical, social, and philosophical implications. While the concept captivates scientific imagination, ethical considerations and responsible use are essential in shaping the discourse around the potential development of superhuman abilities.

4.2.1.1. The Physical and Mental Aspects of Becoming Superhuman

The idea of becoming superhuman involves enhancing both physical and mental aspects of human abilities. Here's a detailed exploration of these aspects:

1. Physical Enhancement: Superhuman physical abilities could encompass increased strength, agility, endurance, and even regenerative capabilities. Bioengineering techniques might target

muscle development, bone density, and cardiovascular efficiency to achieve such enhancements.

2. Cognitive Enhancement: Superhuman mental abilities could involve enhanced memory, processing speed, problem-solving skills, and learning capabilities. Bioengineering might focus on neural connections, neurotransmitter regulation, and brain plasticity to elevate cognitive functions.

3. Ethical Considerations: The pursuit of superhuman capabilities raises ethical questions about fairness, access, and the potential for creating societal divides based on enhanced abilities. Ensuring equitable access to such technologies becomes paramount.

4. Risk and Safety: The safety and long-term consequences of physical and mental enhancements must be rigorously evaluated to avoid unintended health risks and adverse effects on individuals.

5. Identity and Humanity: Philosophical questions emerge about the definition of humanity and the potential impact of superhuman abilities on human identity. Ethical discussions delve into the boundaries of human nature and what it means to be human.

6. Social Impact: Superhuman abilities could influence social dynamics, employment, and the way society values certain skills and attributes. The social implications of these changes warrant careful consideration.

7. Regulatory Framework: Establishing comprehensive regulatory frameworks becomes essential to govern the responsible development and use of bioengineering technologies for human enhancement.

8. Informed Consent: Individuals must provide informed consent when undergoing bioengineering interventions to ensure they understand the potential risks and benefits of becoming superhuman.

9. Enhanced Performance and Competition: In sports and other competitive arenas, the emergence of superhuman abilities raises questions about fairness and the preservation of the spirit of competition.

10. Collaborative Dialogue: Addressing the complex implications of becoming superhuman requires inclusive and collaborative dialogue among scientists, ethicists, policymakers, and the public.

As the potential for human enhancement through bioengineering continues to advance, striking a balance between scientific progress, ethical considerations, and societal impact is paramount. Responsible development and thoughtful engagement are essential in shaping the future of superhuman capabilities, respecting human values, and safeguarding the well-being of individuals and society at large.

4.2.1.2. The Potential Powers and Abilities of Superhumans

The concept of superhumans evokes visions of extraordinary powers and abilities beyond natural human limits. While still in the realm of science fiction and speculative imagination, here's a detailed exploration of some potential powers and abilities that could be associated with superhumans:

1. Enhanced Strength: Superhumans could possess significantly increased physical strength, allowing them to lift heavy objects and perform feats of power that surpass human limitations.

2. Superhuman Speed: Enhanced speed might enable superhumans to move at incredible velocities, making them faster than ordinary humans in both running and reaction times.

3. Enhanced Durability: Superhumans might exhibit enhanced durability, with the ability to withstand physical stresses, extreme temperatures, and even certain types of injuries.

4. Enhanced Senses: Superhuman abilities could include heightened senses, such as enhanced vision, hearing, or olfactory capabilities, providing them with a greater awareness of their environment.

5. Accelerated Healing: Superhumans might possess accelerated healing abilities, allowing them to recover from injuries and illnesses at an extraordinary rate.

6. Enhanced Cognitive Abilities: Superhuman mental abilities could involve enhanced memory, problem-solving skills, and information processing, enabling them to process vast amounts of data quickly.

7. Telepathy or Mind Control: In the realm of science fiction, superhumans might possess telepathic or mind control abilities, enabling them to communicate or manipulate the minds of others.

8. Flight: The power of flight is a classic superhuman ability, granting individuals the ability to defy gravity and soar through the skies.

9. Shape-shifting: Superhumans with shape-shifting abilities could alter their physical appearance, assuming different forms and identities.

10. Immortality or Longevity: In some portrayals, superhumans might possess immortality or significantly extended lifespans, granting them exceptional longevity.

It's essential to emphasize that the concept of superhumans remains speculative and fictional. While bioengineering and scientific advancements offer promising avenues for human enhancement, the realization of such extraordinary powers raises profound ethical, social, and safety considerations. As we explore the possibilities of human enhancement, it is crucial to navigate these questions responsibly and thoughtfully, considering the potential impacts on individuals, society, and the human experience as a whole.

4.2.2. Bionic Organs and Body Renewal Technologies

Bionic organs and body renewal technologies represent cutting-edge advancements in bioengineering. Bionic organs involve the integration of artificial components with human tissue to enhance organ function, while body renewal technologies aim to rejuvenate and repair aging or damaged tissues. These innovations hold great promise in improving human health, extending lifespan, and revolutionizing medical treatments. However, ethical considerations, safety, and accessibility remain essential factors in the responsible development and widespread implementation of these transformative technologies.

4.2.2.1. The Impact of Bionic Organs on the Human Body

Bionic organs, also known as bioengineered or artificial organs, have the potential to revolutionize medicine and significantly impact the human body. Here's a detailed exploration of their impact:

1. Organ Function Enhancement: Bionic organs are designed to replicate or augment the function of natural organs, leading to improved organ function and overall health in individuals with organ deficiencies or failures.

2. Enhanced Biocompatibility: Advancements in materials science and tissue engineering enable the development of bionic organs with improved biocompatibility, reducing the risk of immune rejection and allowing seamless integration with the human body.

3. Increased Longevity and Quality of Life: By providing vital organ support, bionic organs can extend the lifespan and enhance the quality of life for patients suffering from chronic organ diseases.

4. Improved Medical Treatments: Bionic organs offer alternative treatment options for organ failure, reducing the reliance on traditional organ transplantation and addressing the scarcity of donor organs.

5. Potential for Personalized Medicine: Bionic organs can be tailored to individual patients, considering their specific medical needs, genetic makeup, and physiological characteristics.

6. Ethical Considerations: Ethical discussions revolve around questions of access, affordability, and the potential commodification of human body parts.

7. Technological Risks: Despite advancements, bionic organs still carry some risks, such as device malfunctions or infections, requiring ongoing research and safety measures.

8. Regulatory Challenges: The development and implementation of bionic organs require robust regulatory frameworks to ensure their safety, efficacy, and ethical use.

9. Societal Impact: The widespread adoption of bionic organs could influence societal perspectives on disability, organ transplantation, and the boundaries of human capabilities.

10. Future Directions: Ongoing research and innovation in bionic organ technologies continue to expand the possibilities for organ replacement and enhancement.

While bionic organs hold tremendous promise in improving human health and addressing critical medical needs, addressing ethical, safety, and accessibility concerns remains essential. Collaborative efforts among scientists, policymakers, and the medical community are crucial in harnessing the full potential of bionic organs while upholding ethical principles and ensuring equitable access to transformative medical technologies.

4.2.2.2. Bionic Organs and the Lives of Individuals with Disabilities

Bionic organs have the potential to significantly impact the lives of individuals with disabilities, offering new possibilities for improved health and functional capabilities. Here's a detailed exploration of their impact:

1. Enhancing Functional Abilities: Bionic organs can provide individuals with disabilities with enhanced functional abilities, compensating for impairments and enabling them to engage in activities they were previously unable to perform.

2. Improved Independence: By restoring or augmenting organ function, bionic organs can enhance independence in daily activities, reducing the need for external assistance and support.

3. Quality of Life Enhancement: Bionic organs can lead to an improved quality of life for individuals with disabilities, allowing them to participate more fully in social, professional, and recreational aspects of life.

4. Psychological Well-being: Regaining organ function through bionic organs can positively impact the psychological

well-being of individuals, reducing feelings of helplessness and increasing confidence and self-esteem.

5. Reduction of Stigma: Bionic organs can contribute to reducing the stigma associated with disabilities by empowering individuals and changing societal perceptions of disability.

6. Personalized Solutions: Bionic organ technologies can be customized to meet individual needs, providing tailored solutions for specific disabilities and medical conditions.

7. Medical Advancements: The development of bionic organs drives advancements in medical research and technology, benefiting not only individuals with disabilities but the broader medical community.

8. Access and Affordability: Ensuring access to bionic organs for individuals with disabilities requires addressing affordability and equitable distribution of these advanced medical technologies.

9. Ethical Considerations: Ethical discussions revolve around questions of consent, autonomy, and the potential implications of enhancing human abilities.

10. Empowering the Disabled Community: Bionic organs have the potential to empower the disabled community, promoting

inclusivity and equal opportunities for individuals with disabilities.

While bionic organs hold immense promise in transforming the lives of individuals with disabilities, careful consideration of ethical, social, and economic factors is vital. Ensuring affordability, accessibility, and responsible use of bionic organ technologies can maximize their positive impact and contribute to creating a more inclusive and equitable society for individuals with disabilities.

4.2.3. The Social and Ethical Implications of Human-Machine Integration

Human-machine integration raises vital social and ethical questions. Concerns about autonomy, privacy, identity, and inequality must be carefully addressed. Robust regulations and thoughtful ethical considerations are essential to ensure a responsible and inclusive integration that benefits humanity while upholding human values and rights.

4.2.3.1. Human-Machine Synthesis and New Social Dynamics

Human-machine synthesis, the seamless integration of humans and machines, has the potential to reshape social dynamics in profound ways. Here's a detailed exploration of the new social implications:

1. Redefining Human Identity: The fusion of human and machine components challenges conventional notions of human identity, prompting discussions about what it means to be human in an era of enhanced capabilities.

2. Social Acceptance and Stigma: The integration of visible machine components into human bodies may face social acceptance challenges, with individuals potentially experiencing stigma or discrimination.

3. Technological Divide: Unequal access to advanced human-machine integration technologies could lead to a technological divide, exacerbating existing social inequalities.

4. Altered Communication and Interaction: Human-machine synthesis could lead to novel forms of communication and interaction, blurring the lines between humans and technology.

5. Employment and Economic Disruption: Widespread human-machine integration might disrupt traditional employment patterns, necessitating new economic models and workforce adaptations.

6. Trust and Reliability: Building trust in human-machine synthesis requires transparency, reliability, and safeguards against potential risks and errors.

7. Impact on Social Institutions: Social institutions, such as education, healthcare, and governance, may need to adapt to accommodate the changing dynamics of human-machine synthesis.

8. Ethical Decision-Making and Accountability: Ethical considerations become paramount as human-machine interactions involve complex decision-making processes, requiring clear accountability for outcomes.

9. Technological Dependency: Increased reliance on machines could lead to technological dependency, necessitating measures to ensure that individuals retain control and autonomy.

10. Cultural and Artistic Expressions: Human-machine synthesis may influence cultural and artistic expressions, leading to new forms of creativity and cultural narratives.

Navigating the social implications of human-machine synthesis requires proactive engagement among researchers, policymakers, and society at large. Balancing technological advancements with ethical considerations is essential to create an inclusive, equitable, and responsible integration that fosters human well-being and respects human values.

4.2.3.2. The Ethical Implications of Human-Machine Integration on Personal Privacy and Societal Control

The seamless integration of humans and machines raises profound ethical concerns regarding personal privacy and societal control. Here's a detailed exploration of these implications:

1. Privacy Concerns: Human-machine integration involves collecting and processing vast amounts of personal data, raising concerns about data privacy, security breaches, and potential misuse of sensitive information.

2. Surveillance and Monitoring: The integration of machines into human bodies or environments could lead to increased surveillance and monitoring, impacting individuals' autonomy and sense of privacy.

3. Consent and Autonomy: Ethical questions arise about obtaining informed consent and ensuring individuals retain control over their integrated technologies, safeguarding their autonomy.

4. Social Control and Manipulation: Human-machine integration may give rise to new forms of social control and manipulation, affecting individuals' decision-making and behaviors.

5. Discrimination and Bias: Biases in machine algorithms and data inputs may perpetuate discrimination and social inequalities, leading to biased outcomes for certain individuals or groups.

6. Transparency and Accountability: Ensuring transparency in human-machine integration processes is crucial for holding responsible parties accountable for their actions and decisions.

7. Dehumanization and Loss of Identity: Excessive integration of machines into human life may lead to dehumanization and a loss of individual identity, eroding the essence of what it means to be human.

8. Access and Inequality: Unequal access to advanced integration technologies could exacerbate societal inequalities, widening the gap between privileged and marginalized groups.

9. Technological Dependence: Overreliance on integrated technologies might lead to technological dependence, raising questions about human resilience and adaptability.

10. Regulatory Frameworks: The development of robust ethical and legal frameworks is essential to protect personal privacy, ensure informed consent, and address the potential for societal control.

Addressing the ethical implications of human-machine integration requires proactive engagement from researchers, policymakers, and the broader society. Balancing technological advancements with individual rights and societal well-being is paramount in fostering a future where human values and privacy are respected in the face of transformative integration technologies.

CHAPTER 5

Virtual Reality and Digital Imagination

5.1. Virtual Reality and the Perception of Reality

Virtual reality (VR) offers an immersive experience, blurring the boundaries between real and virtual worlds. Users may experience altered perceptions, emotions, and social interactions in VR, raising ethical considerations about its impact on human cognition and well-being. As VR technology evolves, understanding its influence on human perception becomes crucial for responsible integration and harnessing its potential benefits.

5.1.1. The Fundamentals and Development of Virtual Reality Technology

Virtual reality (VR) technology aims to create immersive computer-generated environments that simulate real-world experiences. It involves head-mounted displays (HMDs), tracking sensors, and high-quality graphics to deliver realistic interactions. Over time, VR hardware has become more accessible and diverse, finding applications in gaming, education, healthcare, and more. Ongoing research and development continue to enhance VR technology, paving the way for exciting future prospects and transformative user experiences.

5.1.1.1. The Basic Principles of Virtual Reality Technology

Virtual reality (VR) technology operates on several fundamental principles that enable its immersive and interactive experiences. Here's a detailed overview of these principles:

1. Immersion: VR aims to immerse users in a simulated environment, creating a sense of presence and engagement. This is achieved through visual, auditory, and sometimes haptic stimuli that transport users to the virtual world.

2. Stereoscopic Display: VR utilizes stereoscopic displays in head-mounted devices to present separate images to each eye, creating a three-dimensional perception of the virtual environment.

3. Head Tracking: VR systems employ head tracking sensors to monitor users' head movements in real-time. This tracking enables the VR scene to adjust perspective accordingly, enhancing the feeling of presence and spatial awareness.

4. Motion Tracking: In addition to head tracking, VR often incorporates motion tracking to monitor users' body movements. This allows users to interact with the virtual environment by moving and gesturing.

5. Latency Reduction: Minimizing latency between user input and the system's response is crucial to maintaining a smooth and realistic VR experience, as any delay can lead to motion sickness and decreased immersion.

6. 3D Audio: VR incorporates spatial audio technology to create a realistic auditory experience. Sound sources are positioned in the virtual space to match their corresponding visual objects, enhancing the sense of presence.

7. Interactivity: Interaction is a key aspect of VR. Users can manipulate virtual objects, navigate the environment, and engage with various elements through controllers or gestures.

8. Real-time Rendering: High-performance graphics engines are used to render the virtual environment in real-time, ensuring smooth and responsive visuals as users explore the virtual world.

9. Content Creation: Creating compelling VR content requires careful consideration of user engagement, spatial design, and storytelling techniques to deliver meaningful experiences.

10. Platform and Software Integration: VR technology depends on seamless integration of hardware and software components to provide a cohesive and user-friendly experience.

By leveraging these fundamental principles, VR technology can transport users to diverse virtual environments, enabling novel applications in fields such as gaming, training, education, healthcare, and more. As VR technology continues to evolve, advancements in hardware, tracking, and rendering will further enhance its immersive capabilities and expand its potential applications in various domains.

5.1.1.2. Technological Trends in Virtual Reality

Virtual reality (VR) technology is continually evolving, driven by advancements in hardware, software, and user experience. Here's a detailed exploration of the technological trends shaping the future of VR:

1. Wireless and Standalone VR: The trend towards wireless and standalone VR headsets eliminates the need for tethering to PCs or gaming consoles, enhancing mobility and accessibility.

2. Improved Resolution and Visual Fidelity: Higher-resolution displays and improved optics contribute to sharper and more realistic visuals, enhancing the sense of immersion in VR experiences.

3. Eye and Face Tracking: Integrating eye and face tracking technology allows for more natural interactions within VR environments, enabling gaze-based interactions and enhanced social presence.

4. Varifocal Displays: Varifocal displays dynamically adjust the focus based on users' gaze direction, reducing eye strain and increasing comfort during extended VR sessions.

5. Haptic Feedback and Haptic Suits: Advanced haptic feedback technologies provide users with tactile sensations, such as vibrations and pressure, increasing the level of realism and immersion in VR experiences.

6. Hand and Finger Tracking: Precise hand and finger tracking enable natural hand interactions in virtual worlds, replacing the need for controllers and further enhancing user engagement.

7. Mixed Reality (MR) Integration: The integration of MR technologies allows for seamless blending of virtual and real-world elements, expanding the potential applications of VR in various industries.

8. AI-Powered Environments: AI algorithms enhance VR experiences by creating dynamic and responsive virtual environments, adapting to users' behavior and preferences.

9. Cloud VR Services: Cloud-based VR services enable more extensive and resource-intensive VR experiences without requiring high-end local hardware, making VR accessible to a broader audience.

10. Social VR and Virtual Collaboration: VR platforms are becoming more socially connected, allowing users to interact and collaborate in shared virtual spaces, leading to new possibilities in remote work and social interactions.

11. VR for Training and Education: VR is increasingly used for training and educational purposes, offering safe and immersive environments for skill development and learning.

12. Healthcare and Therapy Applications: VR is proving valuable in healthcare, providing therapeutic solutions for pain management, exposure therapy, and rehabilitation.

As VR technology continues to advance, these trends promise to revolutionize the way we interact with digital content and each other. The seamless integration of cutting-edge hardware, AI, and social connectivity will shape a future where VR becomes an integral part of daily life, offering transformative experiences across diverse domains.

5.1.2. Virtual Reality and Its Impact on Human Perception

Virtual reality (VR) has a profound impact on human perception, blurring the line between reality and the virtual world. By immersing users in computer-generated environments, VR alters perceptions of space, time, and self. It evokes strong emotional responses and can lead individuals to react as if the virtual scenarios were real. As VR technology advances, understanding its influence on human perception becomes essential for responsible integration and maximizing its potential benefits in various applications.

5.1.2.1. The Experience of Virtual Reality and the Degree of Perceived Reality

Virtual reality (VR) technology is designed to create a highly immersive and realistic experience for users, leading to a strong sense of perceived reality in the virtual environment. Several factors contribute to the degree of perceived reality in VR experiences:

1. Immersive Environment: VR uses advanced hardware, such as high-resolution displays and spatial audio systems, to create a multisensory environment that surrounds users and fosters a feeling of presence in the virtual world.

2. Interaction and Agency: VR allows users to interact with and manipulate objects within the virtual environment, providing a sense of agency and control, further reinforcing the illusion of being part of the digital world.

3. Sensory Feedback: Some VR setups incorporate haptic feedback, enabling users to feel tactile sensations, adding to the realism of the experience.

4. Field of View: Wide field-of-view VR displays enhance the sense of immersion by providing a broader and more encompassing view of the virtual world.

5. Latency Reduction: Minimizing latency between user actions and system responses is critical to maintaining the feeling of seamless interaction, preventing any disconnection between the user's movements and the virtual environment.

6. Visual Realism: High-quality graphics and realistic animations contribute to the perceived reality in VR, as lifelike representations of objects and environments create a more convincing virtual experience.

7. Depth and Scale Perception: VR systems accurately render depth and scale, allowing users to perceive distance and object sizes realistically, contributing to a more authentic experience.

8. Emotional Engagement: VR experiences can evoke strong emotional responses, as users may feel emotionally connected to the scenarios they encounter, whether they are thrilling, relaxing, or emotionally impactful.

9. Suspension of Disbelief: Users often suspend their disbelief while in VR, accepting the virtual world as genuine, even though they are aware of its artificial nature.

10. Cognitive Engagement: VR can captivate users' attention and cognitive resources, leading to heightened focus and engagement in the virtual experience.

The degree of perceived reality in VR experiences can vary among individuals and across different VR applications. Factors such as prior VR experience, individual susceptibility to immersion, the quality of the VR hardware, and the content's realism all influence the level of immersion and the sense of presence in the virtual environment.

Understanding the nuances of perceived reality in VR is crucial for content creators and developers, as it influences the potential of VR to elicit profound emotional reactions, facilitate meaningful experiences, and affect user well-being. Striking a balance between immersion and user comfort is essential for crafting impactful and responsible VR experiences that harness the full potential of this transformative technology. As VR technology continues to advance, the exploration of user perception and its ethical implications will be essential in shaping the future of VR applications across various domains.

5.1.2.2. The Mental and Sensory Experience in Virtual Reality

Virtual reality (VR) offers a compelling mental and sensory experience that blurs the line between the real and virtual worlds.

This immersive nature is achieved through a combination of mental and sensory stimuli, creating a profound impact on users' cognitive and perceptual processes:

1. Visual Experience: VR provides a visually captivating experience with high-resolution displays and 3D rendering. Users perceive depth, scale, and spatial relationships, creating a realistic visual environment.

2. Auditory Experience: Spatial audio technology enhances immersion by accurately placing sound sources in the virtual space. This provides users with a sense of direction and distance for audio cues, improving situational awareness.

3. Tactile Sensations: Some VR setups incorporate haptic feedback, allowing users to experience tactile sensations, such as vibrations or pressure, when interacting with virtual objects. This enhances the feeling of presence and realism.

4. Motion and Navigation: Users can move and navigate within the virtual environment, contributing to the sense of agency and control. Motion tracking technologies monitor users' movements, enabling real-time responses.

5. Cognitive Engagement: VR captivates users' attention and cognitive resources, leading to heightened focus and engagement

in the virtual experience. This can enhance learning outcomes and skill development in educational and training applications.

6. Emotional Impact: VR experiences can evoke strong emotional responses, ranging from excitement and joy to fear and empathy. Users may emotionally engage with virtual scenarios, even though they are aware of their artificial nature.

7. Time Perception: The perception of time can be altered in VR, with users reporting a sense of time dilation or compression during immersive experiences. This temporal distortion can impact users' perception of time passing.

8. Sense of Presence: The culmination of visual, auditory, tactile, and interactive elements creates a sense of presence, where users feel as if they are physically present in the virtual world.

9. Suspension of Disbelief: Users often suspend disbelief while in VR, accepting the virtual world as real despite knowing it is a simulation. This suspension of disbelief contributes to the immersive experience.

10. Post-VR Effects: After leaving the virtual environment, some users may experience a period of readjustment, where the lines between virtual and physical reality may temporarily blur.

The mental and sensory experience in VR is a complex interplay of technologies and design elements that work together to create a compelling and transformative experience. As VR technology continues to advance, content creators and developers must consider the psychological impact of VR on users, ensuring responsible use and maximizing its potential benefits in various fields, from entertainment and education to healthcare and training. Understanding the intricacies of the mental and sensory experience in VR will shape the future of immersive technologies and pave the way for new applications and discoveries in human-computer interaction.

5.2. Digital Imagination and Virtual Avatars

Digital imagination and virtual avatars are two interconnected aspects of virtual reality (VR) and digital experiences. Digital imagination refers to the creative and imaginative potential of VR, enabling users to explore and interact with worlds beyond physical constraints. Virtual avatars, on the other hand, are digital representations of users that allow them to immerse themselves in virtual environments and interact with others in a social and dynamic manner. Together, digital imagination and virtual avatars redefine how users perceive themselves and their surroundings, offering a gateway to novel experiences and social interactions in the digital realm.

5.2.1. Digital Imagination Technology and Its Development

Digital imagination technology encompasses the tools and capabilities that enable users to create, modify, and interact with virtual worlds and digital content. It has seen rapid development and expansion in recent years, driven by advancements in computer graphics, artificial intelligence, and interactive design. Digital imagination technology includes virtual reality (VR), augmented reality (AR), and mixed reality (MR) platforms, as well as content creation tools, 3D modeling software, and immersive storytelling techniques. As technology continues to evolve, digital imagination opens up new possibilities for artistic expression, education, training, entertainment, and collaborative experiences in the virtual realm.

5.2.1.1. The Basic Principles and Applications of Imagination Technology

Imagination technology encompasses a set of fundamental principles and versatile applications that empower users to explore and create immersive digital experiences. Here is a detailed exploration of these principles and applications:

1. Virtual Environments: Imagination technology creates virtual environments that replicate real-world settings or imaginative fictional realms. Users can interact with these

environments using VR, AR, or MR devices, offering a sense of presence and immersion.

2. 3D Modeling and Animation: Digital imagination relies on 3D modeling and animation software, enabling the creation of lifelike virtual objects, characters, and landscapes. These tools are crucial for crafting compelling and realistic digital content.

3. Interactive Storytelling: Imagination technology supports interactive storytelling, where users actively participate in narratives, making decisions that influence the storyline. This approach enhances engagement and user agency in digital experiences.

4. Artistic Expression: Artists and designers leverage imagination technology to express their creativity, pushing the boundaries of visual art, sculpture, and digital installations, leading to innovative and thought-provoking creations.

5. Educational Simulations: Imagination technology finds applications in education, providing immersive simulations for training, skill development, and experiential learning. It allows learners to engage with complex concepts in a practical and memorable way.

6. Entertainment and Gaming: Imagination technology is a driving force behind the gaming industry, offering realistic and

captivating gaming experiences. Gamers can venture into fantastical worlds, assume different roles through avatars, and interact with other players in virtual environments.

7. Architectural Visualization: Architects and urban planners use imagination technology to create virtual representations of buildings and urban spaces. It aids in visualizing design concepts, assessing spatial arrangements, and communicating ideas to clients and stakeholders.

8. Virtual Collaboration: Imagination technology facilitates remote collaboration by enabling individuals to meet and work in shared virtual spaces. This is particularly useful for teams spread across different locations.

9. Therapeutic Applications: In healthcare, digital imagination plays a role in therapeutic applications, such as exposure therapy for phobias and anxiety disorders, as well as pain management and rehabilitation.

10. Empathy and Social Impact: Digital imagination can foster empathy by allowing users to experience situations from different perspectives, leading to a greater understanding of diverse experiences and social issues.

As digital imagination technology continues to advance, its applications will diversify further, impacting various aspects of human life and society. It holds the potential to transform how we learn, create, entertain, and communicate, enriching our experiences in both the digital and physical realms. However, along with these opportunities come ethical considerations, such as data privacy, content moderation, and responsible use of technology, which require careful attention to ensure a positive and inclusive digital future.

5.2.1.2. Digital Imagination and Augmented Reality

Digital imagination and augmented reality (AR) are intertwined concepts that enhance our perception of the world by overlaying virtual elements onto the real environment. Here is a detailed exploration of the relationship between digital imagination and augmented reality:

1. Augmented Reality Basics: Augmented reality refers to technology that superimposes computer-generated content, such as images, videos, or 3D models, onto the user's view of the physical world in real-time. This is typically achieved through AR-enabled devices like smartphones, tablets, smart glasses, or AR headsets.

2. Enhancing Real-World Experiences: AR enhances real-world experiences by supplementing the physical environment with digital content, creating an enriched and interactive perception of the surroundings. Users can access information,

instructions, or additional visual elements related to the objects they encounter.

3. Imagination in AR Content: Digital imagination is at the core of AR content creation. AR developers and designers leverage 3D modeling, animation, and interactive storytelling to craft virtual elements that seamlessly blend into the real world, sparking the user's imagination.

4. Immersive Interaction: AR enables users to interact with virtual objects, manipulate them, and even collaborate with others in shared AR experiences. This level of interactivity fosters a sense of presence and engagement, encouraging users to explore and experiment in the mixed reality environment.

5. Applications of AR and Digital Imagination: AR finds applications in various fields, including education, entertainment, retail, marketing, training, and healthcare. Educational AR apps can bring history to life, interactive books can engage young readers, and AR navigation tools can improve wayfinding in unfamiliar places.

6. Social AR Experiences: AR also fosters social interaction, allowing users to engage with one another through shared augmented experiences. Social AR games, collaborative design projects, and virtual meetings in AR spaces exemplify the social potential of this technology.

7. Potential for Creative Expression: AR offers artists and creators a new canvas for creative expression. Through AR art installations and immersive storytelling, artists can merge the physical and virtual realms, evoking unique emotional responses from audiences.

8. Challenges and Considerations: While AR expands our imaginative capabilities, it also raises challenges related to data privacy, digital ethics, and the impact of extended AR use on the physical world. Striking a balance between the virtual and real world experiences is crucial to ensure the responsible adoption of AR technology.

9. Future Prospects: As AR technology advances, the line between digital and physical experiences will continue to blur. The integration of artificial intelligence, spatial mapping, and wearable AR devices will shape the future of digital imagination, offering even more realistic and personalized AR interactions.

The combination of digital imagination and augmented reality opens up vast possibilities for transforming the way we perceive and interact with our environment. It enriches our experiences, sparks creativity, and empowers us to explore new dimensions of the world around us. As we harness the potential of AR and digital imagination, thoughtful design and ethical considerations will be essential in shaping a positive and inclusive augmented future.

5.2.2. Virtual Avatars and Self-Representation in Imagination

Virtual avatars play a significant role in digital imagination, enabling users to represent themselves in virtual environments and experiences. Avatars serve as digital personas, allowing individuals to interact with others, express their identities, and immerse themselves in virtual worlds. The ability to create and customize virtual avatars enhances self-representation, empowering users to project different aspects of themselves and explore new identities within the realm of imagination.

5.2.2.1. Creating Personal Avatars and Personal Representation

Creating personal avatars is a fundamental aspect of digital imagination, offering users the opportunity to design virtual representations of themselves that can be used in various digital experiences. Here is a detailed exploration of creating personal avatars and its significance in personal representation:

1. Customization and Personalization: Users can customize their avatars to reflect their physical appearance, style, and personality traits. This customization allows individuals to tailor their avatars to match their real-world identity or experiment with

new looks and characteristics, giving rise to diverse and unique avatars.

2. Self-Expression and Identity: Avatars serve as an extension of self-expression, allowing users to showcase their interests, hobbies, and preferences. They can choose clothing, accessories, and features that resonate with their identity, enabling them to project their individuality in virtual environments.

3. Empowerment and Inclusivity: Personal avatars empower individuals to represent themselves in digital spaces, irrespective of physical limitations or social barriers. This inclusivity fosters a sense of belonging and participation in virtual communities.

4. Psychological Impact: Creating and using personal avatars can have positive psychological effects, including increased self-esteem and a sense of agency. The ability to embody a digital persona can boost confidence and provide a sense of control in virtual interactions.

5. Social Interaction and Presence: Personal avatars enhance social interactions in virtual environments, as they allow users to engage with others in a more immersive and relatable manner. Avatars contribute to a sense of presence, facilitating meaningful connections and communication.

6. Playfulness and Exploration: Designing personal avatars can be a playful and creative process, encouraging users to explore different possibilities and experiment with their visual representation. This playful element enhances the enjoyment of digital imagination.

7. Cross-Platform Representation: Personal avatars can be used across various digital platforms, including social media, online gaming, virtual meetings, and virtual reality experiences. This versatility ensures consistent self-representation in different digital contexts.

8. Ethical Considerations: While personal avatars offer newfound freedom in self-representation, ethical considerations must be taken into account. Ensuring user consent for avatar usage and addressing concerns related to data privacy are crucial aspects of responsible avatar implementation.

9. Avatar Evolution: As digital imagination technology evolves, avatars are becoming more sophisticated, with advanced facial expressions, body movements, and voice integration. This evolution further enhances the realism and emotional connection with avatars.

10. Future Possibilities: The future of personal avatars holds exciting possibilities, including AI-driven avatars that can dynamically respond to users' emotions and adapt to their

preferences, further blurring the line between digital and physical self-representation.

Creating personal avatars and embracing the notion of personal representation in digital imagination redefines how individuals interact with virtual spaces and how they are perceived within these environments. It promotes a sense of agency, fosters creativity, and reinforces the idea that the virtual world can be a canvas for personal expression and exploration.

5.2.2.2. Avatars in Virtual Communities and Games

Avatars play a central role in virtual communities and games, shaping the user experience and enhancing social interactions. Here is a detailed exploration of avatars in virtual communities and games:

1. Social Interaction and Presence: In virtual communities, avatars serve as the visual representation of users, enabling social interaction in digital spaces. Avatars create a sense of presence, making users feel more connected to one another and fostering a sense of community.

2. Customization and Identity Expression: Virtual community platforms and games offer extensive avatar customization options, allowing users to design unique and personalized representations of themselves. Avatars become a

medium for self-expression, reflecting users' personalities and preferences.

3. Role-Playing and Gameplay: In gaming, avatars often take on specific roles and personas, allowing players to immerse themselves in the game world and adopt different identities. Avatars become the player's alter ego, shaping the gameplay experience and narrative.

4. In-Game Progression: Avatars in games can evolve and progress over time as players achieve milestones and complete challenges. Advancements in avatar appearance or abilities reward players for their accomplishments and provide a sense of achievement.

5. Socialization and Communication: Avatars facilitate communication in virtual communities and multiplayer games. Players can interact with others through their avatars, engaging in conversations, emotes, and non-verbal gestures that mimic real-life social cues.

6. Avatar Diversity and Inclusivity: Virtual communities and games increasingly promote avatar diversity and inclusivity, allowing users to create avatars that represent various ethnicities, genders, and body types. This fosters a more inclusive and welcoming environment for all users.

7. Creative Avatars: Some virtual communities encourage users to showcase their creative skills by designing elaborate and artistic avatars. This creativity adds to the richness of the virtual space and promotes a sense of community engagement.

8. Emotional Attachment: Users often develop emotional attachments to their avatars, considering them extensions of themselves in the virtual world. This emotional connection can lead to a more invested and fulfilling experience within the community or game.

9. Social Avatars and Non-Player Characters (NPCs): In addition to player-controlled avatars, virtual communities and games may feature non-player characters (NPCs) as social avatars. NPCs interact with players, contributing to the narrative and gameplay dynamics.

10. Future Trends: Advancements in virtual reality, augmented reality, and AI will likely lead to even more sophisticated avatars with realistic expressions, voice recognition, and AI-driven behavior. These developments will further enhance the sense of presence and social engagement in virtual communities and games.

Avatars are central to the immersive and social aspects of virtual communities and games, enabling users to forge connections, express themselves, and partake in memorable experiences. As virtual technology continues to evolve, avatars will remain pivotal in shaping the way we interact, communicate, and socialize in the digital realm.

5.3. Digital Imagination and Social Interaction

Digital imagination profoundly impacts social interaction by providing new avenues for communication and collaboration in the digital realm. Through virtual communities, social media platforms, and interactive experiences, digital imagination enriches social interactions, allowing individuals to connect, create, and share in innovative and engaging ways. It fosters a sense of belonging, empowers creative expression, and opens up possibilities for global connections and cultural exchange. As technology continues to evolve, digital imagination will further transform how we interact and build relationships in the digital age.

5.3.1. Digital Imagination and the Role of Social Media

Digital imagination plays a significant role in shaping social media experiences. Social media platforms offer a canvas for creative expression, where users can share personalized content, including images, videos, and virtual art. Avatars and filters further enhance self-representation, allowing individuals to curate their online identity. Digital imagination fuels social media trends, challenges, and interactive features, fostering user engagement and virtual connections. As social media and digital imagination evolve together, they continue to redefine how we communicate, share, and interact in the digital age.

5.3.1.1. Social Media and Self-Identity and Image in Digital Imagination

Social media platforms have become integral to digital imagination, enabling users to construct and present their self-identity and image in virtual spaces. Here is a detailed exploration of the relationship between social media, self-identity, and image in the context of digital imagination:

1. Identity Curation: Social media allows individuals to curate and present their identity in a controlled manner. Users carefully select and share content that aligns with how they wish to be perceived by others, highlighting aspects of their lives, interests, and achievements.

2. Visual Representation: Images and videos are key components of digital imagination on social media. Users can create and share visually captivating content, utilizing filters, effects, and editing tools to enhance their appearance and the overall aesthetic of their posts.

3. Avatar and Profile Customization: Some social media platforms offer avatar customization features, enabling users to

create digital personas that represent their ideal self or specific aspects of their personality. Profile pictures and usernames contribute to the overall online image.

4. Storytelling and Personal Branding: Through posts, stories, and captions, users can craft narratives about themselves and their lives, engaging in personal branding. They build connections with their audience, sharing experiences that resonate and reflect their values.

5. Social Comparison and Identity Exploration: Social media invites users to compare themselves with others, leading to identity exploration and self-reflection. This process can be empowering for some, while for others, it may contribute to self-esteem issues and social pressures.

6. Virtual Communities and Identity Affiliation: Social media fosters the formation of virtual communities based on shared interests, identities, or affiliations. Users find like-minded individuals, contributing to a sense of belonging and validating their self-identity.

7. Impact on Mental Health: Digital imagination on social media can influence users' mental health, as constant self-presentation and social comparison may lead to feelings of inadequacy or anxiety. Being mindful of digital consumption and setting healthy boundaries is crucial.

8. Authenticity and Deception: While social media provides tools for digital imagination, it also raises concerns about authenticity and deception. Users may alter their appearance or present an idealized version of themselves, blurring the line between reality and virtual identity.

9. Social Influence and Trends: Social media influencers and trends can shape digital imagination on a larger scale. Users may adapt their self-identity and image to align with popular trends or follow the examples set by influential figures.

10. Ethical Considerations: Responsible digital imagination involves respecting others' privacy, avoiding harmful content, and being transparent about digital enhancements or filters used in posts. Ethical use of social media ensures a healthier online environment.

In the realm of digital imagination, social media serves as a canvas for self-identity and image expression. Users can construct and share their narratives, connecting with others on personal, creative, and professional levels. While social media facilitates empowerment and communication, it also poses challenges in maintaining authenticity and navigating the impact on mental well-being. Striking a balance between digital self-representation and real-life authenticity is vital for fostering positive and meaningful social interactions in the digital age.

5.3.1.2. Social Media and Its Impact on Virtual Communities

Social media has a profound impact on the development and dynamics of virtual communities, redefining how individuals connect and interact in digital spaces. Here is a detailed exploration of the influence of social media on virtual communities:

1. Global Connectivity: Social media platforms enable virtual communities to transcend geographical boundaries. People with shared interests, regardless of their physical location, can come together, fostering a sense of global connectivity and cultural exchange.

2. Community Formation and Niche Interests: Social media facilitates the formation of virtual communities centered around niche interests, hobbies, or specific topics. Users can find like-minded individuals, contributing to the growth and vibrancy of these communities.

3. Engagement and Interaction: Social media enhances engagement within virtual communities through various interactive features, such as comments, likes, shares, and direct messaging. Users can actively participate in discussions and contribute to community conversations.

4. Community Building and Moderation: Social media provides tools for community organizers to build and moderate their virtual spaces. Admins can create guidelines, moderate content, and encourage positive interactions to maintain a welcoming environment.

5. Amplification of Content: Social media allows community-generated content to reach a broader audience beyond the confines of the virtual community itself. This amplification increases exposure and attracts new members to join the community.

6. Real-Time Communication: Social media enables real-time communication among community members, fostering immediate and dynamic interactions. Live streams, chats, and story features add a sense of immediacy and excitement to community engagement.

7. Collaboration and Co-Creation: Virtual communities on social media platforms often engage in collaborative efforts and co-creation. Members can come together to organize events, projects, or creative endeavors, further strengthening community bonds.

8. Community Identity and Representation: Social media platforms allow virtual communities to establish a distinct identity and representation. Customization of community profiles,

hashtags, and visual elements helps convey the community's essence.

9. Networking and Opportunities: Social media opens doors for networking and opportunities within virtual communities. Members can connect with professionals, potential collaborators, or mentors who share similar interests or career goals.

10. Challenges and Conflict Resolution: While social media enriches virtual communities, it also presents challenges, including managing conflicts and maintaining a positive atmosphere. Community leaders must address issues promptly to preserve harmony.

As a powerful communication tool, social media significantly influences the cohesion, growth, and impact of virtual communities. By fostering connections and facilitating vibrant interactions, social media continues to shape the landscape of digital communities, providing individuals with spaces to share, learn, and thrive in the ever-expanding digital world.

5.3.2. Digital Imagination and Communication Methods

Digital imagination revolutionizes communication methods by introducing innovative and immersive ways for individuals to connect and express themselves. From virtual reality meetings to augmented reality messaging, digital imagination transforms the traditional communication landscape. Users can now engage in richer and more dynamic interactions, leveraging creative tools and virtual environments to communicate ideas, emotions, and stories in ways never before possible. As technology continues to advance, digital imagination will continue to shape the future of communication, providing endless possibilities for human expression and connection.

5.3.2.1. Communication in Augmented Reality and Virtual Worlds

Augmented reality (AR) and virtual worlds offer unique and immersive communication experiences that blend the real and virtual realms. Here is a detailed exploration of communication in augmented reality and virtual worlds:

1. Shared Virtual Spaces: Augmented reality and virtual worlds enable users to interact within shared virtual spaces. Participants can communicate and collaborate as avatars or digital representations, fostering a sense of presence and connection.

2. Real-Time Interaction: Communication in AR and virtual worlds happens in real-time, allowing users to engage with one another instantaneously. This synchronous communication enhances the feeling of being present with others despite physical distances.

3. Non-Verbal Communication: In these digital environments, users can express emotions and intentions through non-verbal cues, such as avatar gestures, facial expressions, and body language. This adds depth and nuance to conversations.

4. Multi-Modal Communication: AR and virtual worlds support multi-modal communication, combining text, voice, and visual elements. Users can use voice chat, text messaging, and virtual objects to convey messages and ideas.

5. Immersive Meetings and Collaboration: Augmented reality facilitates immersive virtual meetings, where participants can interact with 3D objects and data overlays in their physical environment. Virtual worlds enable collaborative experiences where users build and interact with shared digital spaces.

6. Creative Expression: Communication in AR and virtual worlds encourages creative expression. Users can create virtual art, build interactive environments, and craft unique avatars, fostering a sense of individuality and self-expression.

7. Enhanced Learning and Training: AR and virtual worlds offer engaging educational experiences. Learners can participate in interactive simulations, immersive training sessions, and role-playing scenarios, optimizing learning outcomes.

8. Global Connectivity: These communication methods enable global connectivity, connecting people from different corners of the world in shared digital spaces. Cultural exchange and international collaboration become more accessible.

9. Social Networking: Augmented reality and virtual worlds serve as social networking platforms. Users can join communities, attend virtual events, and explore shared interests with like-minded individuals.

10. Challenges and Ethical Considerations: Communication in AR and virtual worlds also poses challenges, such as ensuring digital safety, managing virtual identities, and preventing harassment within digital spaces.

The integration of digital imagination in augmented reality and virtual worlds creates exciting opportunities for communication, collaboration, and creative expression. As technology advances and these platforms become more accessible, communication in these virtual environments will continue to evolve, shaping the way we connect and interact in the digital age.

5.3.2.2. Imagination-Based Communication and Future Directions

Imagination-based communication represents a cutting-edge approach that combines technology, creativity, and human expression to reshape how we interact and connect. As this field continues to evolve, several future directions emerge:

1. Haptic Communication: Future advancements may integrate haptic technology to enable touch-based interactions in virtual environments. Users could feel and transmit tactile sensations, enhancing the sense of presence and emotional connection.

2. Multi-Sensory Experiences: Imagination-based communication will likely encompass multi-sensory experiences, engaging sight, sound, touch, and potentially taste and smell. This convergence of senses will enrich communication, making it more immersive and compelling.

3. Brain-Computer Interfaces: Direct brain-to-computer communication could revolutionize how we express ideas. Imagination-based interfaces could decode mental images and thoughts, enabling communication without the need for verbal or physical expression.

4. AI-Powered Interaction: Artificial intelligence (AI) will play a pivotal role in imagination-based communication. AI algorithms may assist in translating complex ideas into immersive visuals or providing real-time language translation for seamless global communication.

5. Virtual Social Spaces: Virtual social spaces will evolve into sophisticated environments, offering diverse activities, games, and interactive experiences for community-building and fostering relationships.

6. Augmented Empathy: Imagination-based communication may cultivate augmented empathy, allowing individuals to understand and relate to others' emotions more profoundly, fostering deeper connections and mutual understanding.

7. Collaborative Creativity: Future platforms could enable real-time collaborative creation, where users co-create art, music, or immersive environments, encouraging a shared sense of accomplishment and creative expression.

8. Personalized Interaction: Imagination-based communication will become more personalized, adapting to individual preferences and communication styles to create tailored experiences for each user.

9. Ethical Considerations: As imagination-based communication advances, ethical considerations regarding privacy, data security, and digital ownership will require careful attention to ensure user safety and protect digital rights.

10. Integration with Daily Life: Future directions aim to seamlessly integrate imagination-based communication into daily

life, enabling practical applications in education, work, healthcare, and entertainment.

As technology continues to progress, imagination-based communication will undoubtedly play an increasingly prominent role in human interaction. By leveraging the power of human creativity and innovation, this paradigm will shape the future of communication, creating more immersive, engaging, and meaningful connections in the digital realm.

CHAPTER 6

Out-of-Body Experiences

6.1. Digital Brain Uploading and Consciousness Transfer

Digital brain uploading aims to replicate a person's brain in a digital format, potentially preserving memories and thoughts beyond the physical body. Consciousness transfer takes it a step further, transferring the entire consciousness into a new medium. While intriguing, these ideas pose significant ethical and technical challenges, including the nature of consciousness and identity continuity. Achieving these concepts remains a distant possibility, necessitating responsible exploration of their implications.

6.1.1. Digital Brain Uploading and Neuro-Computational Models

Digital brain uploading involves creating a digital replica of a person's brain by employing neuro-computational models. These models attempt to simulate the brain's complex neural connections and functionalities. The data obtained through brain scanning is processed to create a digital representation, which, in theory, could preserve the individual's memories and cognitive processes. While still in the realm of science fiction, advances in neuroscience and artificial intelligence drive research in this intriguing field, holding the potential to revolutionize our understanding of consciousness and personal identity. However, achieving successful brain uploading and ensuring its ethical implementation require overcoming immense technical challenges and addressing complex ethical considerations.

6.1.1.1. Brain Uploading and Brain-Computing

Brain uploading, also known as mind uploading or whole brain emulation, is a speculative concept that involves transferring the entirety of a person's brain, including its neural connections and functions, into a digital or computational substrate. The idea posits that the digital replica would preserve the individual's memories, personality, and consciousness, effectively allowing for the continuation of their existence in a digital form.

The process of brain uploading relies heavily on brain-computing, which involves creating detailed computational models of the brain's neural networks and simulating their activities. To achieve this, sophisticated neuro-computational models are developed, integrating neuroscience, computational science, and artificial intelligence. These models attempt to emulate the complex interactions among neurons and synapses that underlie the brain's cognitive processes and subjective experiences.

The steps involved in brain uploading and brain-computing can be summarized as follows:

Brain Scanning: High-resolution brain scanning techniques, such as advanced MRI or electron microscopy, are used to create a detailed map of the brain's neural architecture.

Data Acquisition: The brain's structural and functional data obtained from scanning are processed and compiled into vast datasets.

Computational Modeling: Neuro-computational models are constructed to simulate the brain's neural activity and the interactions between neurons.

Brain Emulation: The computational model is run on powerful supercomputers or specialized hardware to emulate the brain's activities in a digital form.

Consciousness Transfer: If successful, the digital emulation would represent the individual's consciousness, allowing for the transfer of subjective experiences into the digital substrate.

While the concept of brain uploading captivates the imagination, it remains highly speculative and faces significant challenges:

1. Technological Complexity: The brain is incredibly intricate, with billions of neurons and trillions of connections. Creating a comprehensive and accurate computational model poses enormous technical hurdles.

2. Consciousness and Subjectivity: The nature of consciousness and subjective experience is not fully understood. It is unclear whether replicating neural activity alone can preserve the essence of personal identity and consciousness.

3. Ethics and Identity: Brain uploading raises ethical questions concerning individual identity, personal rights, and consent. The digital replica's legal and moral status remains uncharted territory.

4. Transfer of Experience: Even if brain uploading were successful, the transfer of one's subjective experiences and consciousness to a new digital medium remains speculative and philosophically complex.

Given these challenges, brain uploading and brain-computing are currently in the realm of speculative science fiction rather than established scientific reality. As neuroscience and technology continue to advance, the exploration of these concepts will require careful consideration of their ethical, philosophical, and societal implications.

6.1.1.2. Digital Brain Uploading and the Treatment of Neurological Disorders

Digital brain uploading, the concept of replicating a person's brain in a digital format, holds potential implications for the treatment of neurological disorders. While this idea remains speculative, some speculate that brain uploading could offer new avenues for understanding and addressing neurological conditions:

1. Simulation for Research: A digital replica of a brain could serve as a valuable tool for neuroscience research. Scientists might use the model to study neural networks and test hypotheses about the underlying mechanisms of neurological disorders.

2. Drug Development: Brain uploading could aid in drug development for neurological disorders. Simulated brains might provide a platform for testing potential treatments and predicting their effectiveness before clinical trials.

3. Personalized Medicine: If successful, brain uploading could enable personalized treatments for neurological conditions. Simulated brains might offer insights into an individual's unique neural circuitry, allowing for targeted therapies.

4. Brain Rehabilitation: In theory, a digital brain replica might facilitate rehabilitation efforts for patients with neurological damage. By analyzing the digital model, clinicians could devise rehabilitation plans tailored to the individual's neural characteristics.

5. Brain-Machine Interfaces: Brain uploading concepts intersect with brain-machine interfaces (BMIs). Connecting a

digital brain to external devices could offer novel ways to restore functionality for individuals with neurological disabilities.

It's important to note that these potential applications are currently speculative and depend on significant advancements in neuroscience, computational science, and ethical considerations. Ethical challenges such as consent, privacy, and the definition of personal identity would need careful exploration before any practical applications could be considered.

As of now, brain uploading remains a thought-provoking idea, and while it may inspire new research directions, it is crucial to approach it with scientific rigor and considerate ethical discussions. The treatment of neurological disorders will continue to advance through conventional medical research and technological innovations without relying on brain uploading concepts.

6.1.2. Consciousness Transfer and Digital Existence

Consciousness transfer envisions the relocation of an individual's consciousness from a biological body to a digital or synthetic medium, potentially allowing for digital existence beyond physical limitations. This speculative concept captivates the imagination, presenting possibilities of immortality and exploration of new frontiers.

The idea suggests that by transferring consciousness, one's subjective experience and sense of self could persist beyond the physical body's lifespan. However, the nature of consciousness and its relationship with the brain remain complex and not fully understood, making consciousness transfer currently a subject of philosophical and scientific debate rather than practical application.

Ethical considerations surrounding identity, consent, and the distinction between the original and the replicated consciousness are essential in discussing consciousness transfer. As technology and our understanding of consciousness evolve, responsible exploration and thoughtful discussions will be necessary to navigate the implications of this profound and futuristic concept.

6.1.2.1. The Concept of Consciousness Transfer and Philosophical Challenges

Consciousness transfer, also known as mind uploading or mind copying, is a speculative concept that proposes transferring an individual's consciousness from their biological body to a digital or synthetic substrate. The idea envisions preserving one's subjective experience, memories, and sense of self in a new form, potentially allowing for digital existence beyond the limitations of the physical body. While this concept captures the imagination, it raises profound philosophical challenges:

1. Nature of Consciousness: The concept of consciousness transfer hinges on our understanding of consciousness itself. The nature of subjective experience, self-awareness, and the relationship between the mind and the brain are complex and not fully comprehended. Determining whether a replicated consciousness would be indistinguishable from the original is a fundamental philosophical question.

2. Identity and Personal Continuity: The transfer of consciousness raises questions about personal identity and continuity. If a digital replica were created, would it truly be the same individual with continuous subjective experiences, or merely a copy with its own separate existence? Philosophers debate whether subjective experience is transferable or unique to an individual's physical embodiment.

3. Qualia and Subjectivity: Consciousness involves qualia, the raw, subjective experiences that form our perceptions. Philosophers question whether these qualia could be faithfully transferred, preserved, or even understood in a digital substrate.

4. Self-Location Problem: Consciousness transfer introduces the "self-location problem," which explores how an individual would subjectively experience being in a new digital medium. Would one's sense of self remain in the original body, in the digital replica, or somewhere in between during the transfer process?

5. Dualism and Materialism: Consciousness transfer highlights the ongoing debate between dualism and materialism in philosophy of mind. Dualism posits that consciousness is non-physical and distinct from the brain, while materialism asserts that consciousness arises from physical processes in the brain. The implications of consciousness transfer align closely with these philosophical perspectives.

6. Ethical Considerations: The potential consequences of consciousness transfer raise ethical dilemmas. Issues related to informed consent, personal autonomy, the rights of digital beings, and potential exploitation require careful examination.

While consciousness transfer is currently limited to the realms of science fiction and philosophical inquiry, advances in neuroscience, artificial intelligence, and brain-computer interfaces may prompt further exploration. As this concept captures public interest, philosophical discussions will continue to shape our understanding of consciousness and the profound implications of manipulating subjective experience through technology. Responsible and reflective dialogue is essential as we navigate the philosophical challenges surrounding consciousness transfer and its potential future impact on society and personal identity.

6.1.2.2. Digital Existence and the Possibility of Infinite Life

The concept of digital existence through consciousness transfer raises intriguing speculations about the potential for infinite life. By transferring one's consciousness to a digital medium, proponents envision the possibility of transcending the limitations of the human lifespan and achieving virtual immortality. While an alluring idea, it involves several theoretical considerations:

1. Immortality vs. Continuity: The idea of infinite life through digital existence challenges traditional notions of mortality and the natural life cycle. While a digital entity might persist indefinitely, questions arise about the continuity of personal identity and subjective experience over such extended periods.

2. Technical Feasibility: Achieving digital existence and consciousness transfer requires advanced technologies far beyond our current capabilities. The ability to faithfully replicate an individual's consciousness in a digital substrate remains speculative, and the technological challenges are immense.

3. Ethical Implications: The pursuit of infinite life through digital existence raises ethical questions about the value of finite existence and the potential consequences of seeking immortality.

Issues related to resource allocation, societal impacts, and the meaning of life come to the forefront.

4. Existential Questions: The concept prompts existential reflections on the nature of life, purpose, and the human experience. Would an infinite life in a digital realm offer the same richness and depth as embodied human existence? What would it mean to exist without the limitations of a physical body?

5. The Nature of Consciousness: The transfer of consciousness to a digital medium hinges on our understanding of consciousness and its relationship with the brain. The nature of subjective experience and self-awareness remains a subject of philosophical and scientific inquiry.

As of now, the idea of digital existence and infinite life remains speculative and resides in the realm of science fiction. The possibility of achieving such a state raises profound philosophical, ethical, and societal questions. While technology may continue to advance, the implications of achieving virtual immortality through digital existence warrant thoughtful consideration and responsible exploration to navigate the potential impacts on human society and individual experiences.

6.2. Virtual Living and Ethical Dimensions of Out-of-Body Experiences

Virtual living, where individuals immerse themselves in digital environments through avatars, raises ethical concerns. Out-of-body experiences (OBEs), where users perceive themselves separate from their physical bodies, add complexity. Ethical considerations involve identity, consent, privacy, potential addiction, psychological effects, and societal impacts. Responsible use and safeguarding users' well-being are essential as we explore the ethical dimensions of virtual living and OBEs.

6.2.1. Virtual Living and Digital Societies

Virtual living introduces digital societies where people interact through avatars in virtual environments. This fosters global connections, inclusivity, and virtual economies. Ethical concerns include identity, privacy, and governance. Responsible development is essential for positive impacts on social dynamics and the human experience.

6.2.1.1. Social Norms and Ethical Codes in Virtual Reality

Virtual reality (VR) environments present unique challenges in establishing social norms and ethical codes. As individuals immerse themselves in digital worlds, they interact with others through avatars, creating dynamic social interactions.

To ensure positive and respectful virtual communities, the following considerations are crucial:

1. Respect for Others: Virtual environments should promote respect for others' identities, experiences, and personal boundaries. Harassment, bullying, or discriminatory behavior should not be tolerated.

2. Informed Consent: Users must be informed about data collection, privacy settings, and potential risks in virtual spaces. Transparent consent practices are vital to protect users' rights.

3. Anonymity and Authenticity: Balancing anonymity and authenticity is essential. While some users may prefer anonymity for safety reasons, ensuring accountability and preventing abuse is crucial.

4. Virtual Property and Ownership: Virtual economies may emerge, necessitating ethical guidelines for property rights and digital asset ownership.

5. Digital Persona and Offline Conduct: Users should understand the distinction between their digital personas and real-world identities. Ethical codes should address how virtual actions may impact offline relationships and vice versa.

6. Governance and Moderation: Virtual platforms require effective governance and moderation to enforce ethical standards and maintain a safe environment.

7. Protecting Vulnerable Users: VR communities should consider the needs of vulnerable users, such as children or individuals with mental health challenges, and implement protective measures.

8. Cultural Sensitivity: Virtual communities often bring together diverse cultures and backgrounds. Respect for cultural differences and sensitivities is crucial in promoting inclusivity.

Establishing ethical codes in virtual reality is a collaborative effort involving developers, users, and community members. By fostering a culture of respect, inclusivity, and responsibility, VR can become a platform for positive social experiences and meaningful interactions.

6.2.1.2. Virtual Communities and Their Real-Life Equivalents

Virtual communities, formed through social interactions in digital environments, exhibit both similarities and distinct differences when compared to their real-life counterparts. Understanding these characteristics can shed light on the unique dynamics of each community type:

1. Social Interaction:

Real-Life Equivalents: Real-life communities involve face-to-face interactions, body language, and physical presence, fostering deeper emotional connections and a sense of immediate bonding.

Virtual Communities: In virtual environments, social interactions occur through avatars and digital means like text and voice chat, allowing for connections across geographical boundaries but potentially lacking the depth of real-life interactions.

2. Sense of Belonging:

Real-Life Equivalents: Sense of belonging in real-life communities often comes from shared physical spaces, such as neighborhoods, workplaces, or social clubs.

Virtual Communities: Virtual communities create a sense of belonging through shared interests, common goals, and experiences within the digital realm, transcending physical limitations.

3. Diversity and Inclusivity:

Real-Life Equivalents: Real-life communities are shaped by geographical proximity, leading to a diverse but geographically limited pool of individuals.

Virtual Communities: Virtual environments attract individuals from various regions and cultures, promoting greater diversity and inclusivity through common interests and passions.

4. Communication:

Real-Life Equivalents: Face-to-face communication allows for nuanced interactions, emotional expressions, and immediate feedback.

Virtual Communities: Communication in virtual environments relies on digital means, potentially leading to misunderstandings but also providing a chance for reflection before responding.

5. Shared Activities:

Real-Life Equivalents: Real-life communities engage in physical activities, events, and gatherings that strengthen bonds through shared experiences.

Virtual Communities: Virtual communities share activities in digital spaces, such as multiplayer games, collaborative projects, or virtual events, fostering a sense of camaraderie.

6. Time and Space:

Real-Life Equivalents: Real-life communities operate within the constraints of time and physical distance, affecting the frequency and duration of interactions.

Virtual Communities: Virtual environments facilitate interactions beyond time zones and borders, enabling asynchronous communication and global connections.

7. Identity Exploration:

Real-Life Equivalents: Real-life identities are bound to physical appearances and immediate self-expression.

Virtual Communities: In virtual realms, individuals can experiment with different aspects of their identities through avatars, allowing for creative self-expression and exploration.

8. Access and Reach:

Real-Life Equivalents: Real-life communities provide access to local resources and expertise but may lack exposure to niche interests.

Virtual Communities: Virtual environments offer access to specialized interests and knowledge that may not be available in local real-life communities.

9. Virtual Economies:

Virtual Communities: Some virtual communities develop digital economies around virtual goods and services, creating unique opportunities for entrepreneurship and trade.

10. Ethical Considerations:

Both virtual and real-life communities encounter ethical challenges such as privacy concerns, moderation, and fostering inclusive and respectful environments.

As virtual reality technology advances and digital communities evolve, understanding the similarities and differences between virtual and real-life communities becomes vital. Recognizing the strengths of each community type can foster an integrated approach to building inclusive and meaningful social experiences both in digital and physical spaces.

6.2.2. Ethical Issues and Social Acceptance of Out-of-Body Experiences

Out-of-body experiences (OBEs) raise ethical questions and social acceptance concerns. Validating experiences, respecting beliefs, and ensuring well-being are crucial aspects to address. Open dialogue and ethical considerations foster understanding and acceptance in society.

6.2.2.1. Psychological and Emotional Consequences of Out-of-Body Experiences

Out-of-body experiences (OBEs) can have profound psychological and emotional consequences on individuals who undergo them. Some key aspects to consider include:

1. Altered Sense of Self: During an OBE, individuals may perceive themselves from outside their physical bodies, leading to a sense of detachment from their usual self-identity. This altered self-perception can trigger confusion and existential questioning.

2. Emotional Impact: OBEs can evoke intense emotions such as awe, fear, or euphoria. The extraordinary nature of these experiences may result in a wide range of emotional responses.

3. Integration Difficulty: Integrating the OBE into one's everyday life can be challenging. Individuals may struggle to reconcile the extraordinary event with their regular sense of reality.

4. Coping Mechanisms: The psychological impact of OBEs can vary widely among individuals. Some may develop coping mechanisms, while others may experience distress or feelings of isolation.

5. Spiritual Significance: For some individuals, OBEs hold spiritual or transcendent meanings, which can influence their emotional response and worldview.

6. Therapeutic Potential: In therapeutic contexts, OBEs may be explored to address certain psychological issues or enhance personal growth. Ethical considerations include potential benefits and risks.

7. Supportive Environments: Creating supportive environments and resources for individuals who have had OBEs is essential. Encouraging open discussions and providing access to professional guidance can aid in coping with the consequences.

Understanding the psychological and emotional consequences of OBEs is crucial in offering appropriate support and care for individuals who undergo these experiences. Respectful exploration of their significance and impact can contribute to a more comprehensive understanding of consciousness and the human mind.

6.2.2.2. Society's Perception of Out-of-Body Experiences and Virtual Living

Society's perception of out-of-body experiences (OBEs) and virtual living is a complex interplay of various factors:

1. Cultural Beliefs: Different cultures hold diverse views on OBEs and virtual living, ranging from acceptance as spiritual phenomena to skepticism or even stigmatization.

2. Scientific Perspectives: While scientific interest in OBEs and virtual reality is growing, skepticism remains regarding the objective reality of these experiences and the potential impact on mental well-being.

3. Media and Entertainment: Portrayals of OBEs and virtual living in media and entertainment can influence public perceptions, sometimes sensationalizing or misrepresenting these phenomena.

4. Technological Advancements: As virtual reality technology advances, societal attitudes toward virtual living may evolve, influenced by its increasing integration into various aspects of daily life.

5. Ethical Concerns: Society grapples with ethical questions related to the use of OBEs and virtual reality, such as privacy, consent, and potential addiction or dissociation.

6. Therapeutic Applications: The exploration of OBEs and virtual living for therapeutic purposes raises discussions about their efficacy and potential risks.

7. Personal Experiences: Individuals who have had OBEs or engaged in virtual living play a role in shaping societal perceptions through sharing their experiences and insights.

8. Education and Awareness: Enhancing public awareness and education about OBEs, virtual reality, and their implications fosters a more informed and balanced societal view.

9. Legal and Regulatory Frameworks: Establishing ethical and legal frameworks for research, therapy, and commercial ventures involving OBEs and virtual living ensures responsible practices.

10. Future Impact: As OBEs and virtual living become more prevalent, society's perception may shift, driven by cultural shifts, scientific discoveries, and technological advancements.

Navigating society's perception of OBEs and virtual living requires open dialogue, balanced representations, and a consideration of cultural, scientific, ethical, and individual perspectives. Encouraging responsible and ethical exploration while addressing concerns helps foster a more nuanced understanding of these phenomena in society.

CHAPTER 7

The Future of Ethics

7.1. The Ethics of Transhumanism and Societal Discussions

Transhumanism sparks vital ethical debates on human enhancement, personal autonomy, social justice, and environmental impact. Engaging in transparent and informed discussions is crucial to address the complex ethical implications and create a responsible and equitable future.

7.1.1. The Ethical Foundations of Transhumanism and Human Values

Transhumanism is grounded in ethical principles that prioritize personal autonomy, well-being, social justice, and responsible innovation. These foundations aim to enhance human potential while respecting human dignity and promoting a more equitable and flourishing future for all.

7.1.1.1. Human Freedom and Autonomy Implications

Transhumanism's emphasis on human freedom and autonomy has significant implications for individuals and society:

1.   Enhanced   Self-Determination: Transhumanist technologies offer the potential for individuals to have greater control over their physical and cognitive abilities, allowing them to shape their own identities and destinies.

2. Choice and Personalization: Human enhancement technologies enable individuals to make choices about the modifications they wish to undergo, tailoring enhancements to their unique preferences and needs.

3. Expanding Boundaries: By pushing the boundaries of human capabilities, transhumanism opens up new possibilities for personal growth, exploration, and self-discovery.

4. Ethical Challenges: The pursuit of human freedom and autonomy through transhumanism raises ethical questions regarding the potential for coercion, societal pressures, and unequal access to enhancements.

5. Identity and Authenticity: Ethical considerations involve the preservation of individual identity and the potential impact of enhancements on one's sense of self and authenticity.

6. Responsibility and Agency: With enhanced abilities, individuals must grapple with increased responsibility and ethical decision-making, both for themselves and society at large.

7. Social Implications: Ensuring that transhumanist advancements are accessible to all and do not exacerbate existing social inequalities is a critical consideration in preserving human freedom and autonomy.

8. Privacy and Consent: As technologies become more invasive and intimately connected to individuals, safeguarding privacy and obtaining informed consent become vital ethical concerns.

9. Balancing Individual and Collective Interests: Ethical dilemmas arise in balancing the pursuit of personal freedom with the broader societal impact of transhumanist advancements.

10. Regulating Autonomy: Societal discussions must address the boundaries of autonomy to prevent misuse or harm resulting from unrestricted enhancements.

Transhumanism's ethical commitment to human freedom and autonomy presents both opportunities and challenges. Engaging in thoughtful dialogue and establishing ethical frameworks can guide the responsible use of technology while preserving individual agency and promoting societal well-being.

7.1.1.2. Transhumanism and Social Justice

Transhumanism's relationship with social justice encompasses several key aspects:

1. Equitable Access to Enhancements: Transhumanism advocates for equal access to transformative technologies, striving to ensure that enhancements do not become exclusive to privileged individuals or groups.

2. Addressing Health Disparities: Transhumanist innovations in medicine and biotechnology have the potential to address health disparities by providing treatments and enhancements to those with medical needs.

3. Enhancing Disabilities: Ethical discussions within transhumanism explore the responsible use of technology to enhance the lives of individuals with disabilities, offering adaptive technologies and assistive devices.

4. Inclusivity and Diversity: Transhumanism values the inclusion and representation of diverse perspectives, recognizing that technological developments should address the needs and aspirations of all members of society.

5. Balancing Individual Aspirations and Collective Well-being: Ethical considerations involve balancing individual desires for enhancement with the collective well-being and ensuring societal cohesion.

6. Environmental Impact: Social justice discussions within transhumanism encompass the environmental consequences of

technological advancements, seeking sustainable and eco-friendly approaches.

7. Ethical Governance: To uphold social justice, transhumanism emphasizes the establishment of ethical governance and policies, preventing the misuse of technology and safeguarding against discriminatory practices.

8. Redefining Norms: Transhumanism challenges traditional norms and stereotypes, aiming to create a more inclusive society that embraces human diversity and differences.

9. Empowerment and Agency: By promoting individual empowerment and autonomy, transhumanism seeks to enable individuals to make informed choices about their own bodies and lives.

10. Human Rights and Dignity: Transhumanism upholds human rights and dignity, recognizing the inherent value and worth of every individual, regardless of their physical or cognitive abilities.

Societal discussions on transhumanism and social justice explore how technology can be harnessed to create a more equitable and inclusive future. Ethical considerations are essential to ensure that advancements align with social justice principles, benefiting humanity as a whole and fostering a more just and compassionate society.

7.1.2. Technological Ethics and the Responsibility of Artificial Intelligence

Ethical considerations in AI development involve ensuring fairness, transparency, and accountability, while safeguarding privacy and promoting human-AI collaboration. Building awareness and establishing ethical guidelines are vital to foster responsible AI innovation that benefits humanity and aligns with human values.

7.1.2.1. The Role of Ethics in Decision-Making Processes of Artificial Intelligence

Ethics plays a crucial role in shaping the decision-making processes of artificial intelligence (AI). Key aspects include:

1. Value Alignment: Ensuring AI systems are aligned with human values and moral principles, guiding their actions and decisions in ways that prioritize human well-being.

2. Bias Mitigation: Addressing biases in AI algorithms to prevent discriminatory outcomes and promote fairness in decision-making.

3. Ethical Frameworks: Establishing ethical frameworks that guide AI systems in complex situations, considering the consequences of actions on individuals and society.

4. Transparency and Explainability: Enhancing AI systems' transparency to understand how decisions are reached, enabling users to trust and comprehend the reasoning behind AI actions.

5. Moral Dilemmas: Ethical considerations involve preparing AI to navigate moral dilemmas, making choices that prioritize human safety and dignity.

6. Impact Assessment: Evaluating the potential societal impact of AI decisions, especially in critical domains like healthcare, finance, and criminal justice.

7. Human Oversight: Implementing mechanisms for human oversight and intervention in AI decision-making to prevent undesirable outcomes.

8. Responsibility Attribution: Defining accountability and responsibility for AI actions, especially in cases where harm may occur.

9. Long-term Implications: Ethical considerations extend to considering the long-term societal consequences of AI decisions, including potential ethical, social, and economic ramifications.

10. Constant Learning and Adaptation: AI systems must be designed to continuously learn and adapt ethical norms and values, reflecting evolving human understanding and ethical standards.

Ethics in AI decision-making is an ongoing area of research and development, emphasizing responsible AI deployment and ensuring that AI technologies serve humanity's best interests while adhering to ethical principles.

7.1.2.2. The Ethical Implications of Artificial Intelligence and Human Control

The ethical implications of artificial intelligence and human control are multifaceted and include the following aspects:

1. Autonomy and Agency: Ethical discussions revolve around the balance of AI autonomy and human agency in decision-making processes to avoid undue concentration of power in AI systems.

2. Accountability and Responsibility: Determining who is accountable for AI actions, especially in cases where AI operates autonomously, raises concerns about responsibility attribution and potential consequences.

3. Human Oversight and Intervention: Ethical guidelines advocate for the integration of human oversight and intervention

mechanisms to ensure that AI decisions align with human values and priorities.

4. Avoiding Unintended Consequences: Ensuring that AI systems do not produce unintended or harmful outcomes due to a lack of human control is a central ethical concern.

5. Preventing Bias and Discrimination: Ethical considerations involve preventing AI systems from perpetuating societal biases and discrimination, which may occur when human biases are encoded in AI algorithms.

6. Preserving Human Dignity: Ethical implications center around preserving human dignity in AI applications, particularly in sensitive areas such as healthcare, education, and law enforcement.

7. Transparency and Explainability: The ethical imperative of AI systems being transparent and explainable enables humans to understand and question the reasoning behind AI decisions.

8. Ethical Decision-Making Frameworks: Establishing ethical decision-making frameworks in AI development guides the incorporation of moral principles into AI systems' actions.

9. Consent and Informed Decision-Making: Ethical concerns arise when AI systems make decisions that significantly impact individuals without their informed consent or understanding.

10. Ethical Standards and Regulation: Ethical implications underscore the importance of developing robust ethical standards and regulations that govern AI development and deployment to protect individuals and society.

Balancing human control and AI autonomy while upholding ethical principles is essential to build trust and ensure the responsible integration of AI into various domains. Ethical considerations are crucial in defining the extent of AI's role and ensuring that human values and ethics remain at the forefront of AI advancement.

7.2. Concerns of Inequality and Social Justice

Transhumanism and technological advancements raise concerns about unequal access to enhancements, widening socio-economic disparities, health inequalities, and potential job displacement. Ethical governance and equitable practices are vital to address these issues and ensure social justice in the integration of advanced technologies.

7.2.1. Technology and Social Inequalities

Technological advancements can exacerbate social inequalities through unequal access to technology, disparities in education, healthcare, and job opportunities. Addressing these challenges requires a focus on equitable practices and policies to ensure technology benefits all segments of society.

7.2.1.1. The Digital Divide and Accessibility Barriers

The digital divide refers to the gap between individuals and communities with access to information and communication technologies (ICTs) and those without. It encompasses various accessibility barriers that hinder marginalized groups from fully participating in the digital world. Some key points to consider include:

1. Infrastructure Disparities: Unequal distribution of ICT infrastructure, such as broadband internet, restricts access for individuals in rural or economically disadvantaged areas.

2. Affordability: High costs of digital devices and internet services can be prohibitive for low-income individuals and families, limiting their participation in the digital ecosystem.

3. Digital Literacy: Lack of digital literacy skills and knowledge prevents some individuals, particularly older generations and marginalized communities, from effectively using technology.

4. Educational Inequalities: Students from underprivileged backgrounds may lack access to computers and the internet, hindering their educational opportunities and digital learning experiences.

5. Employment and Economic Barriers: Limited access to digital tools and online job platforms can impede job-seeking and career growth for disadvantaged individuals.

6. Healthcare Access: Limited access to digital health services and telemedicine can result in reduced healthcare options for remote and vulnerable populations.

7. Social Isolation: Inability to connect digitally may lead to social isolation for individuals without access, especially during periods of remote work and online social interactions.

8. Participation in Civic and Political Activities: Limited digital access can hinder civic engagement and the ability to access important government services and information.

9. Cybersecurity Concerns: Individuals without access to secure digital networks are vulnerable to cyber threats and data breaches, further widening inequalities.

10. Bridging the Gap: To address the digital divide and accessibility barriers, initiatives focusing on digital inclusion, affordability, and digital literacy training are essential. Public and private partnerships can play a vital role in ensuring equitable access to technology for all.

Bridging the digital divide is crucial for creating an inclusive and equitable society, where everyone can benefit from the opportunities presented by digital technologies. It requires a concerted effort from governments, organizations, and communities to reduce accessibility barriers and promote digital literacy and inclusion.

7.2.1.2. Technology and Economic Inequalities

Technological advancements have significant implications for economic inequalities:

1. Job Displacement: Automation and AI can lead to job displacement, affecting low-skilled workers and widening income gaps.

2. Skills Gap: The rapid pace of technological change may create a skills gap, where certain individuals lack the necessary expertise to participate in emerging industries.

3. Wealth Concentration: Advancements in technology can concentrate wealth in the hands of a few tech giants, exacerbating income inequality.

4. Access to Opportunities: Limited access to technology and digital resources can hinder economic opportunities for disadvantaged communities.

5. Gig Economy: The rise of the gig economy may lead to precarious work conditions and reduced job security for vulnerable workers.

6. Educational Disparities: Unequal access to digital learning tools can impact educational outcomes, perpetuating economic disparities.

7. Digital Entrepreneurship: Lack of access to technology and funding can impede entrepreneurship opportunities for marginalized entrepreneurs.

8. Financial Technology: While fintech can increase financial inclusion, some individuals may lack access to digital banking services, limiting their financial growth.

9. Economic Mobility: Unequal access to digital resources may hinder economic mobility and perpetuate intergenerational poverty.

10. Addressing Inequalities: Policymakers and organizations need to address economic inequalities by investing in digital infrastructure, promoting digital literacy, and ensuring technology benefits all segments of society.

Creating a more inclusive digital economy requires targeted efforts to bridge the economic gap and provide equitable access to technology, education, and economic opportunities for all individuals, regardless of their socio-economic background.

7.2.2. The Social and Economic Effects of Transhumanism

Transhumanism's impact on society and the economy is complex. It may lead to social integration challenges, job market transformations, enhanced productivity, and improved healthcare. Ethical considerations, questions about human identity, and access to enhancements also play crucial roles in shaping its effects. Striking a balance between innovation and responsible regulation is essential in managing its implications.

7.2.2.1. Superhumans and Social Segregation

The emergence of superhumans through transhumanist technologies raises concerns about social segregation and discrimination:

1. Enhanced vs. Non-Enhanced: The existence of individuals with enhanced physical or cognitive abilities could create a societal divide between those who have access to enhancements and those who do not.

2. Social Hierarchy: Superhumans may be perceived as superior, leading to the establishment of a new social hierarchy based on technological enhancements.

3. Access and Affordability: Transhumanist technologies might initially be expensive and only accessible to the wealthy, deepening existing economic disparities.

4. Employment Discrimination: Employers may favor superhuman candidates, leading to discrimination against non-enhanced individuals in the job market.

5. Identity and Belonging: Non-enhanced individuals might feel marginalized or inadequately valued, affecting their sense of identity and belonging in society.

6. Social Exclusion: The fear of being left behind or excluded from the benefits of transhumanist technologies could lead to social tension and division.

7. Ethical Concerns: Debates surrounding the fairness of human enhancements and the potential for creating an "us vs. them" mentality must be addressed.

8. Legal Protections: Legal frameworks may need to be revised to safeguard the rights and dignity of both enhanced and non-enhanced individuals.

9. Societal Harmony: Striking a balance between embracing innovation and preventing social segregation is crucial for maintaining societal harmony.

To mitigate the risks of social segregation, policymakers, ethicists, and technology developers must collaborate to ensure equitable access to transhumanist technologies and foster a culture of inclusivity and respect for individual differences. By addressing these concerns, society can harness the potential of transhumanism for collective progress while upholding ethical principles and social cohesion.

7.2.2.2. Genetic Modification and Social Status

The prospect of genetic modification raises significant implications for social status and inequality:

1. Genetic Privilege: Those with access to genetic enhancements may gain a genetic advantage, leading to a new form of genetic privilege and social stratification.

2. Educational Divide: Genetic modifications that enhance cognitive abilities may widen the educational divide between those who can afford enhancements and those who cannot.

3. Health Disparities: Genetic enhancements could exacerbate health disparities by creating unequal access to preventive measures and disease treatments.

4. Employment Bias: Employers may favor genetically enhanced candidates, creating bias and discrimination against non-enhanced individuals in the job market.

5. Reproduction and Family Dynamics: The ability to genetically modify offspring may impact family dynamics and reinforce notions of genetic superiority.

6. Ethical Concerns: Debates about the ethics of germline editing and designer babies are critical in addressing concerns related to social status and genetic inequality.

7. Human Identity: The acceptance and normalization of genetic modifications may influence perceptions of what it means to be human and how we define human worth.

8. Societal Values: Societal values around genetic enhancements may shift, influencing the way we perceive and treat individuals based on their genetic makeup.

9. Legal and Policy Frameworks: The development of responsible legal and policy frameworks is essential to address the social implications of genetic modification.

10. Public Perception: Public perception of genetic modifications and their impact on social status may shape public attitudes and acceptance.

Addressing the social implications of genetic modification requires thoughtful deliberation, public discourse, and ethical considerations. Ensuring equitable access to genetic technologies and upholding human rights will be crucial in fostering a society that values inclusivity, fairness, and respect for all individuals, regardless of their genetic makeup.

7.3. Ethical Frameworks for the Future

Future ethical frameworks must prioritize human dignity, inclusivity, responsibility, and public engagement. They should address potential risks, balance innovation with caution, and encourage international collaboration. Education and regular reevaluation are crucial in shaping a future that harnesses technology for the greater good while upholding ethical values.

7.3.1. Transhumanism and Humanity's Vision for the Future

Transhumanism represents humanity's vision of a future where technological advancements enable us to transcend our current limitations, achieve greater potential, and improve the human condition. Ethical considerations in transhumanism must align with our collective goal of creating a society that values human dignity, fosters inclusivity, and ensures responsible use of technology. As we navigate the possibilities of transhumanism, it is essential to embrace innovation while keeping in mind the social, cultural, and ethical implications. By actively engaging in ethical discussions and collaborative decision-making, we can shape a future where technology and humanity coexist harmoniously, serving the betterment of all.

7.3.1.1. Humanity's Evolutionary Future and Transhumanist Horizons

Transhumanism envisions a future where humanity evolves through advanced technologies, unlocking new horizons and possibilities:

1. Biological Enhancement: Transhumanists explore the potential of genetic modifications, regenerative medicine, and biotechnologies to enhance human physical and cognitive capabilities.

2. Cognitive Augmentation: Brain-computer interfaces and neurotechnologies aim to augment human intelligence, memory, and learning abilities.

3. Immortality and Longevity: Transhumanists seek ways to extend human lifespan and even achieve biological immortality through advancements in medical technologies.

4. Virtual Existence: Digital brain uploading and virtual reality may offer opportunities for conscious existence beyond physical bodies.

5. Post-Humanism: Some transhumanists contemplate a post-human future, where humanity integrates with technology to become a new species with enhanced capabilities.

6. Ethical Explorations: Alongside technological advancements, ethical discussions arise, addressing issues of identity, privacy, social equality, and responsibility.

7. Collective Decision-making: Humanity's evolutionary future demands collective decision-making, including diverse perspectives, public discourse, and collaboration among stakeholders.

8. Balancing Advancements and Values: Striking a balance between technological advancements and human values becomes crucial in shaping the future.

9. Social and Cultural Impacts: Transhumanist horizons require understanding the broader social and cultural impacts of these advancements on human societies.

10. Co-creation of the Future: Humanity's evolutionary future involves active co-creation, where individuals, societies, and institutions collectively shape the path ahead.

Transhumanism raises profound questions about the nature of humanity, the boundaries of existence, and the potential for transformative changes in society. It challenges us to confront ethical dilemmas while embracing a future that integrates technology responsibly and aligns with the values that define our shared humanity. As we venture into this uncharted territory, it is essential to remain conscious of our moral compass, ensuring that our evolutionary journey remains grounded in the principles that honor the essence of being human.

7.3.1.2. Transhumanism and Space Exploration

Transhumanism and space exploration share a symbiotic relationship, envisioning a future where technological advancements propel humanity beyond Earth:

1. Human Adaptation: Transhumanist technologies could aid in human adaptation to the challenging conditions of space, enhancing resilience and survivability.

2. Long-Distance Travel: Genetic modifications and life-extension technologies may be vital for sustaining astronauts during extended space missions.

3. Cognitive Enhancement: Cognitive augmentation could improve decision-making, problem-solving, and creativity during space exploration.

4. Teleoperation and Robotics: Transhumanist concepts, such as brain-machine interfaces, might facilitate remote control of robots and devices in space exploration.

5. Human-Machine Symbiosis: Space missions could involve human-machine symbiosis, combining human ingenuity with AI and advanced robotics.

6. Interstellar Travel: Transhumanist aspirations for immortality and longevity align with the ambitious goal of interstellar travel in the distant future.

7. Colonization Efforts: Space colonization may rely on genetic adaptations and biological enhancements to thrive in extraterrestrial environments.

8. Post-Human Spacefarers: As space missions extend beyond our solar system, discussions on post-human spacefarers may arise, imagining beings with augmented abilities for deep space travel.

9. Ethical Considerations: Space exploration raises ethical questions regarding the preservation of diverse life forms and the potential impact on extraterrestrial environments.

10. Vision for Humanity's Future: Transhumanism and space exploration present a shared vision of humanity's evolutionary journey, expanding our boundaries and striving for a multi-planetary existence.

Both transhumanism and space exploration challenge us to embrace technological progress while confronting ethical dilemmas. By integrating responsible technology and ethical frameworks, we can navigate the vast cosmos, fostering a future that combines our transformative potential with the wonder and curiosity that drive us to explore the unknown frontiers of space.

7.3.2. The Ethics of the Future and Society's Responsibilities

In the face of rapid technological progress, society bears the crucial responsibility of prioritizing human values, promoting inclusivity, and ensuring transparent governance. Ethical education, global collaboration, and continuous evaluation will guide us in navigating a future that upholds human dignity and serves the collective well-being. By embracing responsible innovation, we can shape a future that harmonizes technology with our ethical principles, fostering a thriving and equitable society for generations to come.

7.3.2.1. Future Generations and Sustainability

In contemplating the ethics of the future, it is crucial to consider the impact of our decisions on future generations and the sustainability of our actions. Key considerations include:

Long-Term Consequences: Ensuring that our choices today do not compromise the well-being and opportunities of future generations.

Environmental Stewardship: Implementing sustainable practices to protect the environment and preserve natural resources for the coming generations.

Intergenerational Justice: Striving for fairness between current and future generations, acknowledging the long-term effects of our actions.

Responsible Resource Management: Using resources wisely and responsibly, avoiding depletion that would burden future inhabitants.

Education and Awareness: Fostering awareness among current generations about the importance of ethical decision-making for the future.

Technological Legacy: Considering the ethical implications of leaving a technological legacy that positively impacts the lives of future generations.

Societal Resilience: Building resilient societies that can adapt and thrive in the face of future challenges.

Ethical Innovation: Promoting technological advancements that align with ethical principles, benefitting future generations.

Ethical Inheritance: Transmitting ethical values and principles to successive generations, fostering a culture of ethical responsibility.

Collective Accountability: Recognizing the collective accountability of society in shaping a sustainable and ethical future.

By embracing sustainability and considering the needs and rights of future generations, we can establish a solid ethical foundation for our actions and choices today, ensuring a better and more equitable world for those who come after us.

7.3.2.2. Society's Decision-Making Processes for the Future

As society confronts complex ethical challenges in shaping the future, robust decision-making processes are essential to ensure responsible and inclusive outcomes. Key aspects include:

Inclusivity: Encouraging diverse perspectives and representation in decision-making to avoid bias and promote fairness.

Public Engagement: Involving the public in discussions and deliberations on ethical issues to reflect societal values.

Scientific Input: Seeking guidance from experts and scientific research to inform ethical considerations.

Policy Formation: Developing clear and comprehensive policies to guide ethical decision-making across various domains.

Ethical Committees: Establishing independent ethics committees to evaluate and advise on emerging technologies.

Transparency: Maintaining transparency in decision-making processes to build public trust and accountability.

Long-Term Vision: Considering the long-term implications of decisions, especially in relation to future generations.

Adaptability: Embracing flexibility to adapt ethical frameworks as technology and societal dynamics evolve.

Global Cooperation: Engaging in international cooperation to address global ethical challenges collectively.

Ethical Education: Promoting ethical literacy and awareness to empower individuals in making ethical choices.

Value Reflection: Regularly reflecting on ethical values and their application in shaping the future.

By establishing inclusive, transparent, and informed decision-making processes, society can navigate the complexities of the future responsibly, safeguarding human values and building a sustainable and equitable world for generations to come.

Conclusion

"Transhumanism and the Future of Humanity: Infinite Potential" explores the captivating realm of transhumanism, delving into the fascinating interplay between technology, biology, and human existence. From genetic modifications and brain-computer interfaces to virtual living and consciousness transfer, the book unearths the boundless possibilities that await humanity.

Throughout its pages, the book navigates through the profound implications of transhumanism, not merely as a theoretical concept but as a potential reality on the horizon. It scrutinizes the ethical dimensions, social impact, and responsibilities associated with wielding such transformative power.

As technology accelerates, the societal and ethical landscape evolves. This book serves as a compass, guiding readers to ponder the intricate questions surrounding human enhancement, digital existence, and the path we traverse into the future.

With thoughtful analysis and comprehensive research, "Transhumanism and the Future of Humanity" invites readers to embark on a thought-provoking journey—a journey that transcends the boundaries of today's human experience and ventures into the uncharted territory of infinite potential.

In the face of these transformative possibilities, this book serves as a reminder of the importance of responsible innovation, inclusive decision-making, and the preservation of human values. As we chart the course to a future where humans and technology intertwine, may we walk this path with empathy, wisdom, and a shared commitment to create a world that benefits all of humanity.

Ultimately, the book serves as a testament to the boundless human imagination and the indomitable spirit of progress. As we venture into the unknown, let us embrace the challenges and seize the opportunities, knowing that the future of humanity lies in our hands, shaped by the choices we make today.

"Transhumanism and the Future of Humanity: Infinite Potential" invites you to explore a world where the line between human and machine blurs, where possibilities become limitless, and where the convergence of humanity and technology holds the key to a bold and transformative tomorrow.

www.ingramcontent.com/pod-product-compliance
Lightning Source LLC
Chambersburg PA
CBHW072138290526
45794CB00004B/1367